TRANSPORTATION INFRASTRUCTURE - ROADS, HIGHWAYS, BRIDGES, AIRPORTS AND MASS TRANSIT

HISTORIC COVERED BRIDGE SECURITY

TRANSPORTATION INFRASTRUCTURE - ROADS, HIGHWAYS, BRIDGES, AIRPORTS AND MASS TRANSIT

Additional books in this series can be found on Nova's website under the Series tab.

Additional e-books in this series can be found on Nova's website under the e-book tab.

TRANSPORTATION INFRASTRUCTURE - ROADS, HIGHWAYS, BRIDGES, AIRPORTS AND MASS TRANSIT

HISTORIC COVERED BRIDGE SECURITY

BENJAMIN DONNELLY
EDITOR

New York

For permission to use material from this book please contact us:
Telephone 631-231-7269; Fax 631-231-8175
Web Site: http://www.novapublishers.com

Library of Congress Cataloging-in-Publication Data

ISBN: 978-1-63117-041-6

Published by Nova Science Publishers, Inc. † New York

CONTENTS

PREFACE

Covered bridges used to cover the American landscape from coast to coast. However, because of various factors including neglect, arson, vandalism, and natural disasters, currently less than 700 to 900 bridges are still in existence today. This book discusses the covered bridge security manual and the remote monitoring of historic covered timber bridges for the prevention of arson and vandalism.

Chapter 1 – The design, construction, and use of covered timber bridges is all but a lost art in these days of pre-stressed concrete, high-performance steel, and the significant growth both in the volume and size of vehicles. Furthermore, many of the existing covered timber bridges are preserved only because of their status on the National Registry of Historic Places or the diligent maintenance and care of the owners of these structures, or both. Of the covered timber bridges that remain in the United States, only a small percentage still stand today because of arson, vandalism, neglect, natural disasters, and other factors. The objective of this work is to provide covered timber bridge owners with the tools to quickly and efficiently design and implement a security system to protect these important historical landmarks. This goal was obtained with an in-depth analysis of equipment based on the practicality in a covered bridge application. Other major considerations required for all equipment to work efficiently within a security system are also fully discussed, including, but not limited to, maintenance, power requirements, and general set-up of an integrated security system. A comprehensive case study is presented involving monitoring systems placed on five of the six covered bridges in Madison County, Iowa, at the end of this report to showcase the abilities of an integrated system and all the decisions that must be made throughout the process for the entire system to work as intended.

Chapter 2 – More than 700 historic covered timber bridges remain in the United States. In recent years, a number of these bridges have been targeted by vandals and arsonists. Obviously, prevention of future acts of vandalism and arson is important to retaining the invaluable parts of our engineering heritage.

A research team from Iowa State University, working cooperatively with the United States Department of Agriculture Forest Products Laboratory (FPL), developed a remote security and fire detection monitoring system for covered timber bridges. This system automatically notifies emergency response personnel in the event that a fire is detected or if unauthorized activity occurs. The system was installed on the Cedar Bridge in Madison County, Iowa in 2005.

Three independent detection systems were software-integrated and installed. Fiber optic sensors were installed to monitor ambient temperatures throughout the bridge. An infrared camera, which is able to detect temperatures in an electronic image, was installed near the bridge. The data from these two parts of the monitoring system are used to detect rapid temperature changes due to fire and to also identify unauthorized persons on site. In addition, an industrial flame detector, which monitors radiant emissions within infrared and ultraviolet wavelengths, was mounted within the bridge interior. All of these components were connected using an onsite, wireless computer network with custom developed system integration software.

In: Historic Covered Bridge Security
Editor: Benjamin Donnelly
ISBN: 978-1-63117-041-6

Chapter 1

COVERED BRIDGE SECURITY MANUAL*

Brent Phares, Terry Wipf, Ryan Sievers and Travis Hosteng

ABSTRACT

The design, construction, and use of covered timber bridges is all but a lost art in these days of pre-stressed concrete, high-performance steel, and the significant growth both in the volume and size of vehicles. Furthermore, many of the existing covered timber bridges are preserved only because of their status on the National Registry of Historic Places or the diligent maintenance and care of the owners of these structures, or both. Of the covered timber bridges that remain in the United States, only a small percentage still stand today because of arson, vandalism, neglect, natural disasters, and other factors. The objective of this work is to provide covered timber bridge owners with the tools to quickly and efficiently design and implement a security system to protect these important historical landmarks. This goal was obtained with an in-depth analysis of equipment based on the practicality in a covered bridge application. Other major considerations required for all equipment to work efficiently within a security system are also fully discussed, including, but not limited to, maintenance, power requirements, and general set-up of an integrated security system. A comprehensive case

* This is an edited, reformatted and augmented version of U.S. Department of Agriculture, Forest Service, Forest Products Laboratory publication, General Technical Report FPL-GTR-223, dated May 2013.

study is presented involving monitoring systems placed on five of the six covered bridges in Madison County, Iowa, at the end of this report to showcase the abilities of an integrated system and all the decisions that must be made throughout the process for the entire system to work as intended.

This study is part of the Research, Technology and Education portion of the **National Historic Covered Bridge Preservation** (NHCBP) Program administered by the Federal Highway Administration. The NHCBP program includes preservation, rehabilitation and restoration of covered bridges that are listed or are eligible for listing on the National Register of Historic Places; research for better means of restoring, and protecting these bridges; development of educational aids; and technology transfer to disseminate information on covered bridges in order to preserve the Nation's cultural heritage.

This study is conducted under a joint agreement between the Federal Highway Administration–Turner Fairbank Highway Research Center, and the Forest Service – Forest Products Laboratory.

INTRODUCTION

Covered bridges used to cover the American landscape from coast to coast. However, because of various factors including neglect, arson, vandalism, and natural disasters, currently less than 700 to 900 bridges are still in existence today. Covered bridges tend to be in isolated areas, and they are constructed with a flammable material, which makes them highly susceptible to arson and vandalism. Because they are usually far from populated areas, it is difficult for firefighters to arrive at the bridges in a timely manner once a fire has been set. This makes it likely that any fire may cause critical damage or may completely destroy a bridge. Most of these covered bridges are covered by volunteer fire departments, which only adds another level of complexity to the situation. It is essential that we protect and preserve these standing landmarks throughout our country for future generations. One way to provide the necessary security is to use remote monitoring and other security systems to protect them from trespassers.

Multiple bridge owners throughout the United States have taken measures to ensure the safety and integrity of their covered bridges by implementing structural strengthening, fireproofing, and monitoring systems. Common protective methods entail illuminating the bridge site and using fire- retardant material around important structural members. These methods are typically used because they are usually very cost effective. Unfortunately, these systems are passive approaches and provide no mechanisms to alert the proper authorities and increase the chances of survival for the bridges.

This report is intended to assist covered bridge owners in selecting systems that will protect covered bridges with the option of adding active alerts to proper authorities for the purpose of heightening security. Because very little technical literature exists dealing exclusively with the subject of security and monitoring systems for covered bridges, we contacted multiple Departments of Transportation (DOTs) with a high concentration of covered bridges as well as multiple bridge owners to assess the state of covered bridge security in the United States. This insight from bridge owners coupled with technical literature about physical security of other assets is the basis of this report.

Overview—Covered Bridge Surveillance Project, Iowa State University

In 2005, the Bridge Engineering Center (BEC) at Iowa State University (ISU) completed a case study on remote security of historic covered bridges by designing and implementing an active monitoring system on the Cedar Covered Bridge in Madison County, Iowa (Phares 2006). The security system installed on the Cedar Bridge is discussed further in the case studies portion of this report. The work completed and the system developed for the Cedar Bridge resulted in the BEC receiving a grant from the Federal Highway Administration (FHWA), through the National Historic Covered Bridge Preservation Program (Matias 2011), to install remotely monitored security systems on the other five covered bridges in Madison County. The five bridges included in the security monitoring grant, the Cutler–Donahoe, Holliwell, Hogback, Imes, and Roseman covered bridges, are five of the six landmark bridges that remain in Madison County and the only bridges that have not been completely rebuilt. The Hogback Bridge was partially damaged by arson but managed to survive and was restored to its original condition one year after the incident. The Cedar Bridge, made famous by the book, *The Bridges of*

Madison County by Robert James Waller, was completely destroyed by arson in 2002 but rebuilt to original specification in 2003–2004 (Overington 2003).

For the Cutler–Donahoe, Holliwell, Hogback, Imes, and Roseman bridges, we decided to design a basic system with a high degree of functionality that could be tailored to fit the majority of covered bridge sites encountered rather than to design a security system to be specific to a given bridge site. The developed security system includes two flame detectors, an infrared camera, an optical camera, and multiple pieces of communication equipment necessary to relay all the information back to the design team as well as to officials in Madison County. We were required to install an alternative energy system at the Hogback Bridge site because it is so distant from local grid power. An in-depth analysis of the security system for these five bridges and problems encountered can be seen later in this report.

The implementations in Madison County were not completed to show what every covered bridge monitoring system should entail. Rather, they provided a mechanism to showcase different ideas, constraints, and limitations for designing a surveillance system for other bridges. This manual is intended to showcase the most predominant and functional technology in the area of surveillance and monitoring at the time of its preparation.

Scope of Work

This manual is intended to provide covered bridge owners with a tool to aid in the development and deployment of security strategies. The research team has completed an extensive government literature review at the local, state, and Federal levels, which included contacting several DOTs and local governments with a high concentration of historic covered bridges that have installed, or are planning to install, security systems or other preventative measures on their bridges. Multiple companies that specialized in outdoor security systems were contacted to define cutting-edge technologies in the security industry at the time this report was written. Note that this manual is not intended to be an all-inclusive list of possible security system equipment and tactics, but a guide to best practices, techniques, and equipment.

LITERATURE REVIEW

Pertinent Covered Bridge Literature

Very little information related to the security of covered bridges has been published. As a result, the majority of the information summarized here deals with general security of bridges or general asset security.

Although there is a lack of published literature dealing exclusively with covered bridges, numerous professionals throughout the country have substantial knowledge related to covered bridges. When possible, these professionals were contacted to provide additional information.

The Beginning of Covered Bridges in the United States

The first covered bridges in the United States appeared in the Eastern states in the early 1800s and thereafter were constructed across the country until the early 1900s when steel bridges became a more economical choice for bridge construction. Covered bridges built during this period were built to reflect the architectural style of a specific time and place while still being a functional and vital passageway for the community. Numerous builders and architectural styles created unique structures that personified communities around the country.

As years passed, the covered bridge became a romanticized icon of the United States of the 1800s. As bridge and road design became routine and repetitive, the distinctive architecture of the past started to stand out more and more when compared with its modern counterpart. Although new construction of covered bridges ceased around the turn of the 20th century, they were still used well into the new century and many continue to be renovated and repaired for use today and well into the future. Despite the best efforts to preserve the bridges at the turn of the 20th century, many covered bridges were lost to neglect, fire, flooding, and other disasters both natural and manmade (Becker 2011).

Different statistics vary about how many covered bridges were constructed and how many still stand today. However, it is clear that a relatively small percentage still stand today when compared with the turn of the 20th century. Recently, the Federal government as well as multiple state and local governments have placed significant emphasis upon preserving our covered bridges. This is evident by the allocations in governmental budgets for spending on covered bridge preservation. Many bridge owners are currently installing security systems to provide protection against arson and vandalism or plan to do so.

Structural Integrity

The bridges were covered for several different reasons. Many historians believe it was to give the appearance that the bridge was a barn and this would have a calming effect on farm animals as they crossed waterways that would otherwise startle them. People also believe that that the bridges were covered to protect travelers who were caught in inclement weather or to hide people on romantic strolls. These ideas may have been considered by the builders when putting coverings on their bridges; however, the main reason that bridges were covered was for a more practical and functional reason. Designers felt that by covering the heavy and expensive trusses from direct rainfall and sunlight, the life expectancy of the bridge could be extended. The designers were correct; history has proven that a covered bridge will last up to three times longer than a similar non-covered bridge. Although the sacrificial wall and roof coverings would have to be replaced every couple of decades, it was still a more economical choice than completely replacing the bridge or structural trusses in the same time frame. The ability to extend the life of timber truss bridges by using sacrificial wall and roof coverings was discovered early in the age of bridge design in Europe and other parts of the world and was an integral part of timber bridge concepts when it arrived in North America (Becker 2011).

Targeting Covered Bridges: Arson and Vandalism

Experts estimate that over 176,000 intentional outdoor fires are set by arsonists every year. These fires result in approximately 20 deaths, 250 injuries, and $23 million in losses annually according to Volume 9, Issue 6 of the Topical Fire Report Series (National Fire Data Center 2009). Out of all the outdoor fires that occur every year in the United States, 27% are intentionally set by arsonists (National Fire Data Center 2009). Arson is a prevalent problem in the United States that destroys property and life and must be addressed in a serious manner. Intentionally set outdoor fires tend to be more common in the spring from March and April and once again in mid-summer, especially July 3–5, according to the National Fire Incident Reporting System

(NFIRS). Because these are the times when arson is most likely to occur, it is recommended that, at a minimum, the highest level of security be active during these time periods.

Table 1. Bridge fires (1992–2002)

Year	Bridge name	Comments
1992	Loy's Station	
1992	Parker	Survived
1993	Slate Bridge	
1993	Jordan	
1993	Corbin	
1993	Smith	
1993	LeMay Ferry	
1993	Nectar C.B.	
1993	Sells	
1993	Kilgore Mill	
1994	Wolf Bridge	
1994	Grimes	
1994	Guilford	
1994	Kaufman's Distillery	
1994	Upper Sheffield	
1995	Miller Road	Survived
1996	Wimer	Survived
1996	Carman	
1997	Offult Ford	
1997	Lower	
1997	Wilkinson	
2000	Henniger Farm	Survived
2001	Pine Grove	Survived
2002	Ryot Bridge	
2002	Orne Bridge	
2002	River Road	
2002	Cedar Bridge	
2002	Henderson	Survived
2002	Risser's Mill	
2002	Jackson	Survived
2002	Jeffries Ford	
2002	Woodsville	Survived
2002	Newfield	Survived
2002	Wilson's Mill[a]	Survived

[a]Wilson’s Mill Covered Bridge (Avella, Pennsylvania) is thought to have survived because of a metal deck.

The devastating effects of arson are evident when committed on covered bridges in small towns all over the country. Unfortunately, this kind of damage is prevalent throughout recent history as seen in Indiana (Rinehart 2005), Iowa (Overington 2003), and Pennsylvania (Murphy 2008). The number of covered bridges throughout the United States is quickly dwindling because of arson as well as neglect. On average, over the last 20 to 30 years, two to three bridges apparently have been set on fire with one or two of these bridges being completely destroyed, as shown in Table 1. It is essential that cities and states that own and maintain covered bridges take the proper measures to ensure that these bridges will survive for future generations to enjoy.

Indiana

In 2005, a fire destroyed one of the 31 covered bridges remaining in Park County, Indiana. The Bridgeton Bridge was beloved by all in the area and was known as the most photographed bridge in the county before the fire. A 35-year-old male, who was a person of interest in an arson case with another covered bridge in the area, seemingly poured an accelerant across the length of the bridge and ignited it sometime around midnight. Both of the bridges were completely destroyed and collapsed. A few months prior to these arson cases, firefighters were able to save another bridge, which also appeared to be intentionally set on fire (Rinehart 2005).

Iowa

Madison County, Iowa, is well known for its covered bridges because of the popularity of the 1992 book, *The Bridges of Madison County,* and the 1995 movie adaption with leading actors Clint Eastwood, Meryl Streep, and Annie Corley. None of the six remaining covered bridges in the county were made as popular by the book as the Cedar Bridge, which is on the cover of the book and is a central bridge throughout the entire story. On September 3, 2002, the Cedar Bridge was completely destroyed by arson, only 4 years after over $128,000 was invested in restoration. The town was extremely distraught over the loss of the bridge, given the large amount of tourism it brought to the community and the manner in which it was destroyed. This prompted the city to replace the structure with an exact replica with as similar of construction techniques as possible the next year (Overington 2003).

Pennsylvania

One of the three remaining bridges in Erie County, Pennsylvania, was set ablaze in December 2008 by two local men. The destruction of this particular bridge was damaging to the city for two separate reasons. The historic Gudgeonville Covered Bride in Girard Township was valued at over $1 million and was of great importance to local tourism and also a vital part of the roadway system that still carried traffic over the Elk Creek. Both of the men responsible for these crimes were suspects in other crimes throughout the area including burglary, criminal trespass, and theft that are all unrelated to the bridge incident (Murphy 2008).

Although arson is a major problem throughout the United States, especially when it pertains to covered bridges, several steps can be taken to reduce the chances of trespassers committing destructive acts toward the covered bridge. In this report, we discuss several successfully implemented recommended procedures for covered bridge security around the United States. With care and diligence, the probability that arson and vandalism will occur on a particular bridge can decrease significantly.

Importance of Historical Integrity

It cannot be stressed enough that any modifications completed on covered bridges including security systems or any other type of rehabilitation must be completed with the greatest of care to ensure that the historical significance of the covered bridge is preserved. The National Register of Historic Places has strict guidelines as to what will and will not be accepted as a historical place. To be eligible for some funds from government agencies to preserve a covered bridge, it is important that a covered bridge is on the register of historic places before and after any modifications. The National Register of Historic Places has strict criteria that must be met for all historic places. It is important that all bridge owners be aware of these criteria before embarking on any project in, on, or around a covered bridge.

Listing in the National Register of Historic Places provides formal recognition of a property's historical, architectural, or archeological significance based on national standards used by every state. Benefits include the following:

- Becoming part of the National Register Archives, a public, searchable database that provides a wealth of research information.
- Encouraging preservation of historic resources by documenting a property's historic significance.
- Providing opportunities for specific preservation incentives, such as

    - Federal preservation grants for planning and rehabilitation
    - Federal investment tax credits
    - Preservation easements to nonprofit organizations
    - International Building Code fire and life safety code alternatives
- Possible state tax benefit and grant opportunities.
- Involvement from the Advisory Council on Historic Preservation when a Federal agency project may affect historic property.

Physical Security of Structures

Physical security is a basic principle necessary for the survival of any person, place, or object. Throughout history, people have used security to protect their privacy, property, and lives whether it be with a weapon, by constructing large impenetrable walls, or more recently by using cameras or other monitoring devices to detect and apprehend ill-willed individuals. By definition, security means the freedom from danger, fear, or anxiety (Merriam Webster Dictionary). By this definition alone, security covers a wide array of areas including information, physical, political, and monetary, as well anything that requires protection from danger. To some degree, there is a concept of security dealing with almost all areas of life, which makes it difficult to cover security in depth at all levels.

Structures can come under attack from terrorism, sabotage, natural disasters, and other threats that may be unique to a particular area. It is essential when designing a security system for a certain structure that a full analysis is done to ensure that the level of security required is obtained. Designing a security system for any structure is always a case of planning for unknown dangers in type and magnitude, with the understanding that not all security risks and dangers can be prevented, deterred, or even detected.

Numerous security measures can be taken for all types of structures whether it be a bridge, building, or any other type of asset. Many of these types of security will not transfer from one type of structure to another. A building, for example, may be under a high level of security by only allowing traffic through approved entrances and restricting all unwanted entry through physical barriers or alarm systems or both. Although the ability to completely monitor all traffic going through and around the structure is ideal for almost all applications, it is not necessarily the most feasible. This is especially true for bridges, which are designed to allow traffic to flow easily. Security measures that require stopping and checking all modes of transportation using the bridge would prove to be uneconomical and unreasonable. However, many precautions and security measures may be implemented to greatly increase the security of the structure and decrease the probability of dangerous behavior occurring.

Blue Ribbon Panel Workshop on Bridge and Tunnel Security

In 2003, the FHWA organized a Blue Ribbon Panel (BRP) Workshop on Bridge Security Assessments following the terrorist attacks of September 11, 2001, to analyze the safety of the transportation infrastructure (DOT 2003). The BRP was given the objective to "Develop short- and long-term strategies for improving the safety and security of the nations' bridges and tunnels, and provide guidance to highway infrastructure owners/operators." The BRP decided upon five major levels of security to construct an effective defense against unwanted activities on or toward bridges. These levels include 1) deter, 2) deny access, 3) detect presence, 4) defend the facility, or 5) design structural hardening to minimize consequences to an accepted level. The BRP report goes into detail about designing structural hardening but does not go into much detail about the other four areas.

Although all levels of security are not equally discussed in full detail within the BRP report, the research team felt it was prudent to investigate and discuss all five levels of security in greater detail. By considering all five levels of security when developing a security system for a bridge or any other structure, the owner is able to assemble a system that is not only effective, but redundant, and that provides the greatest level of security possible for the given budget. Within this report, the concept of structural hardening is covered only briefly as several different publications look at structural hardening of bridges and covered bridges in particular. This report's main goal is to complement these publications with the other four main portions of security discussed within the BRP's assessment. Below, these four levels are briefly discussed; furthermore, multiple examples of each of the areas are discussed within the equipment options portion of this report.

Deter

Deterrence is prevention or discouragement of a detrimental action by means of fear or doubt. For the case of bridge security, the fear created would be that anyone who would commit detrimental actions to the bridge would be caught and prosecuted. One of the more inexpensive options for deterrence would be to place signs around the bridge, alerting all visitors that surveillance equipment and alarms are implemented on site, and local authorities will be alerted of any trespassers after hours. Quite simply, the presence of this type of sign creates fear or doubt (whether or not the surveillance equipment actually exists or not).

Greater levels of deterrence can be implemented if the designer of the security system believes that the bridge is more susceptible to arson or vandalism. For example, in extreme situations, deterrence by means of onsite security personnel may be warranted. Additional levels of increased security may include lighting, barricades, or alarm systems placed such as to frighten any trespassers off the premises. In some cases, the goal of these measures is, quite simply, to create a sense of uncertainty and fear about a particular site.

Deny

Denying trespassers access to the bridge site is one of the more difficult portions of the security program laid out by the BRP. Because most covered bridges are in secluded areas, trespassers may enter the site several different ways. Many of the methods used for deterrence can also be able to deny trespassers to a limited degree. The use of fences and barricades around the bridge area could deny, or at least slow down, potential threats to the bridge.

Detect

In the case of arson and vandalism, surveillance of the bridges is a great first step in the detection. An added benefit is that, once detected, other steps can be taken to lessen the probability of total destruction of the bridge, as will be discussed further. Much of the equipment discussed in this report pertains to the detection of threats. The detection of threats can be an invaluable portion of many security systems. This detection of threats is important because it facilitates alerting local authorities if there is anything going on at the bridge site.

Defend

The ability for the covered bridge to defend itself once a fire has been started can be essential for the survival of the bridge. There are multiple ways to defend a bridge such as sprinkler systems or the application of fire-retardant paint or wood during construction or renovation. Defense is fundamentally different from structural strengthening, as the addition of defensive measures does not change the bridge in a structural sense but only protects the existing structure from fire.

Education

Although all of these physical security measures are important to ensure the highest level of security at a covered bridge, the importance of education should never be overlooked when discussing preventative measures against arson and vandalism. By educating the public in areas surrounding a covered bridge about its historical significance and the importance of preserving the structure, awareness of the bridge is increased and those who take interest in a sense become added security to the bridge. This can easily be done by placing articles in local newspapers, going to local middle and high schools to discuss the history of covered bridges, posting signs, and many other options.

COVERED BRIDGE MONITORING SYSTEM: DESIGN

Preliminary Decisions

Multiple decisions must be made to deploy an effective security system that can fulfill the needs and desires of each bridge owner. Some of these main decisions are discussed in detail within this section; however, these decisions and discussions may not be all inclusive, and proper discretion should be used when implementing a covered bridge security system. Illustrated in Figure 1 are some of the more major decisions that go into designing a security system.

Subsequent sections provide more detail regarding the information and terms presented in Figure 1 to give the designer a more detail for each part of the system.

All equipment or components that are used on or around the bridge site must be carefully selected so that they do not detract from the aesthetic value of the bridge and are able to survive in a hostile environment, all the while providing acceptable levels of performance with minimum maintenance. These criteria prove to be the most important aspects of any equipment choice for most covered bridge applications. All components of the security system that are located on the bridge site should be weather resistant, resistant to the constant abuse of insects or animals, and be placed so that the components themselves are protected from vandalism. An in-depth look measures the need to protect equipment and is discussed in further detail in the set-up section of this report.

All the following equipment explanations are designed to be a brief overview and not a complete narrative. Multiple sources exist such as online reviews and other publications that can assist the security system designer when selecting equipment options based upon their abilities and limitations. Detailed communication with all equipment manufacturers can ensure that all devices can be integrated effectively and meet or exceed the requirements of the bridge owner.

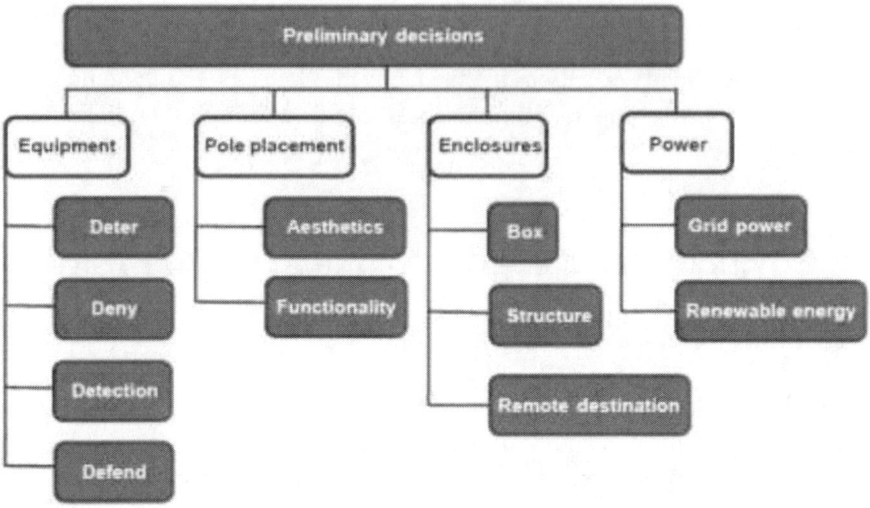

Figure 1. Equipment options.

Equipment Options

Each covered bridge is unique in its aesthetics, structural design, and surrounding landscape, so equipment may be listed here that would not be applicable to a specific bridge. The designer of the security system for a particular bridge should be sensible and rational about what pieces of equipment are selected. In later sections of this report, we discuss which systems might be best used for different locations and situations.

None of the equipment listed below is intended to be used stand-alone to protect a structure. Several pieces of equipment that work together to protect the bridge need to be integrated as efficiently as possible. The five areas of security that are listed by the BRP (i.e., deter, deny, detect, defend, and strengthen structure) cannot be obtained with a single piece of equipment. Any one piece of equipment may only provide one or possibly two types of security; therefore, to obtain optimal levels of security, it is essential to use multiple pieces of equipment that have multiple abilities, as shown in Figure 2.

A decision-making tool, shown in Figure 3, was created to aid bridge owners in making quick decisions about the type of security system to be installed on a particular bridge with a given budget. All prices are considered to be average and may be significantly higher or lower depending on the abilities of the equipment and the specific manufacturer. Each piece of equipment that is listed in this spreadsheet is expanded upon in this section with capabilities and limitations of all equipment.

Deterrence Equipment

Alarm System—Any of the above pieces of security equipment, either stand-alone or together in a system, are nothing more than hardware without the integration of a carefully designed alarm system. Chapter 104, Tyska and Fennelley (2000) discusses alarm systems in great detail. Several different alarm systems may be used depending on the intent and level of security. The different types of systems can include alarms that silently send out an alarm to the proper authorities, a system that emits an audible alarm, some type of visual alarm in the area of the bridge, or a combination of the two. Several different issues may factor into the decision on what type of alarm system to use, such as the bridge's proximity to local authorities and residential areas. If the bridge is in close proximity to a residential area, it may not be attractive to have a loud, audible alarm that could disturb residents for all alarms, real and false. The proximity to the local authorities, and more specifically their response time, is also a key factor in the decision of the type of alarm system. If the bridge is within a reasonable distance to a fire department or police station, it may be more appropriate to have a silent alarm that will only alert the local authorities and give them the possibility of apprehending the criminal. If there is more considerable distance between the bridge and the local authorities or if the bridge is protected by a volunteer fire department, an audible alarm may alert both the criminal and nearby local public with the intent of deterring the criminal before more damage is inflicted to the structure.

A standard alarm system generally consists of the following:

- Detection elements located at the protected area, designed to initiate alarm upon entry of an intruder.
- Transmission lines that conduct signals to a device in the immediate area or to a central annunciator panel that can be continuously monitored.

- A panel that announces by visible or audible signals the structure or area in which an alarm has been activated.
- Fail-safe features that provide a signal at the annunciator panel if any part of the system is malfunctioning.

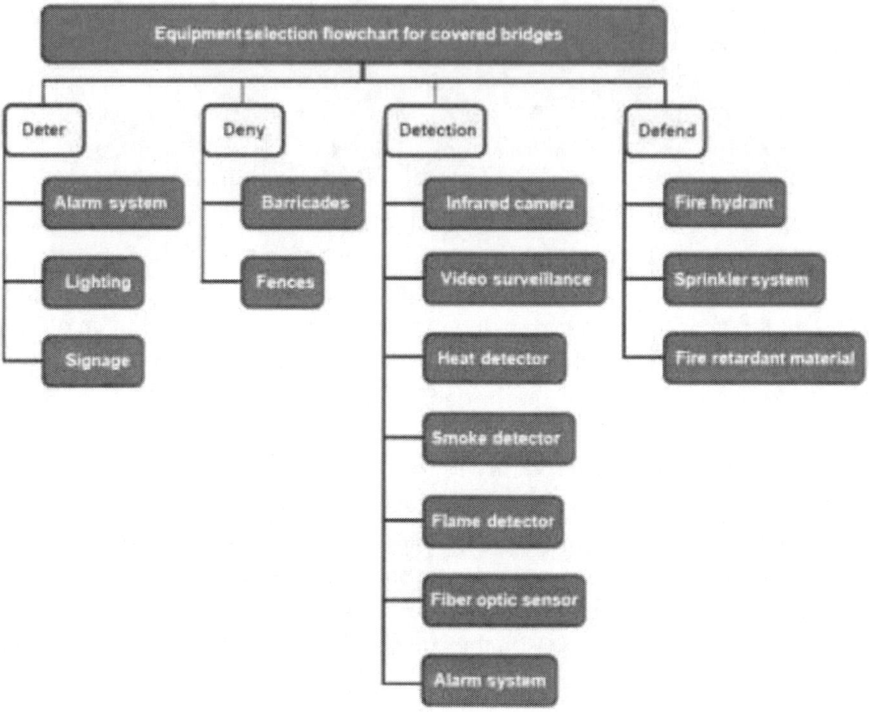

Figure 2. Flow chart of equipment options.

Lighting—Most vandalism and arson are committed during the nighttime hours because the trespassers are disguised by the darkness. Adequate lighting at the bridge site at all times can prove to a cost effective and efficient way of protecting the bridge. In Chapter 24, Tyska and Fennelley (2000) discuss lighting and note the importance of using the right amount of lighting that is effective yet does not contribute to any light pollution, which could be unattractive and costly. Many guidelines can be used to limit light pollution on the bridge site including the use of sensors, timers, high-efficiency fixtures, directional fixtures, and others. It may also be more economical in the long run for a professional with experience with exterior lighting to design the lighting system. Proper lighting can be used in unison with an optical camera to improve effectiveness. Multiple types of lighting can be used in a variety of situations. They include the following:

Perimeter Lighting: This type of lighting is used to illuminate the fence or perimeter of a certain site. The perimeter for covered bridge sites could be at a certain distance from the bridge or be the perimeter of the bridge itself. Perimeter lighting is used so that trespassers must pass through an adequately lit perimeter that may or may not be under surveillance. This is to act as a deterrent to the trespassers.

Area Lighting: This type of lighting is used to illuminate the area immediately surrounding the bridge that must be passed through to enter the bridge. Much like perimeter lighting, area lighting is used as a deterrent to trespassers by making it necessary for people to pass through brightly lit areas to reach the bridge.

Floodlighting: Floodlighting is used to saturate an entire area with bright lighting to deter trespassers. This type of lighting may detract from the aesthetics of the bridge and distract people driving by and living near the bridge site.

Gatehouse Lighting: This type of lighting illuminates entrances and exits to and from the site. This is essential for the entrances of both sides of the bridge as well as entrances to major roadways around the bridge site.

Signage—Placing signage around the entrances of the bridge site stating that anyone who enters after nightfall or attempts to deface the bridge will be prosecuted to the fullest extent of the law is an effective way of deterring trespassers. Signage stating that there is a surveillance system on site that is running 24 hours a day should also be considered for added deterrence. The more signage that is present on the site, the more likely it is that people entering the site will see it. In Chapter 89, Tyska and Fennelley (2000) discuss the usage of signage in great detail.

Deny Equipment

Barricades: In Chapter 4 of *Perimeter Security*, Michael Arata (2006) discusses barricades in full detail. Many covered bridges today are closed to motor vehicles to reduce live loads on the bridge. Barricades offer a variety of advantageous features. Foremost is that they make it difficult for trespassers to get too close to the bridge with motor vehicles. This is important, as this means that they must travel distances on foot, which may slow entry to the site, and more importantly, exit after a defacement has occurred. There are several different barrier options including natural and man-made styles.

	Equipment	Average Price	****Threat Level	Average Capabilities	Main Drawbacks	Requirements	Maintenance
[illegible]	Alarm System		MA/MV	Audible/Visible signals at structure Notify proper authorities	Possibility of false alarms at bridge site Can notify trespasser and allow escape	Low electrical draw Enclosure for communication	Monthly
	Annunciator Panel	$300					
	Lighting	$70/fixture	MA/MV	Make portions of bridge site visible day/night Allow cameras to see during low light levels	Can lead to light pollution	Moderate electrical draw	Yearly
	Signage	[illegible]	LA/LV	Inform trespassers of consequences			Rare
[illegible]	Barricades	[illegible]	LA/MV	Limit traffic to certain sized vehicles Stop traffic altogether from entering bridge	May be difficult to move when necessary		Rare
	Decorative Planters	$300					Rare
	Bollards	$150					Rare
	Fences	$15/foot	LA/MV	Limit access points at bridge site	Cannot completely limit access points		Rare
	Chain Link	$8/foot					Rare
	Wood	$10/foot			Rot and warping of wood may occur over time		Yearly
	Wrought Iron	$25/foot					Rare
	Brick Wall	**$10/ft²					Rare
[illegible]	Infrared Camera	$5,000	HA/LV	Monitor temperatures within field of view View bridge regardless of amount of light	False alarms due to glare, animals, or heat Sensitive to environment	High electrical draw Pole	Yearly
	Video Surveillance	$300	LA/MV	View bridge during times of adequate light Allow remote parties to view bridge	Dependent on lighting in field of view	High electrical draw Pole	Yearly
	Heat Detector	$130	HA	Monitor temperatures around detector	False alarms due to environment	Moderate electrical draw	Yearly
	Smoke Detector	[illegible]	MA	Monitor smoke levels around detector	Highly vulnerable to exterior conditions	Low electrical draw	Monthly
	Flame Detector	$1,500	HA/LV	Monitor temperatures within field of view	False alarms due to glare, animals, or heat	Moderate electrical draw	Yearly
	Fiber Optic Sensor		MA	Monitor temperatures along length of device	Possibly difficult with installing	Moderate electrical draw	Yearly
[illegible]	Fire Hydrant	**$3,000	MA	Provide water source to fire department	Requires close water supply	Close water supply	Yearly
	Sprinkler System	**$11,000	HA	Extinguish fires started within interior of bridge	Requires close water supply Bulky and not aesthetically pleasing	Close water supply	Yearly
	Fire Retardant Material	[illegible]/ft²	HA	Stop or slow the ignition of a fire Stop the spread of any fires that may start	Loss of effectiveness over time Loss of structural strength to applied members		Every few years
	Intumescent Coating	[illegible]/ft²					
	Fire Retardant Wood	***%20 increase					

Notes:

* Draw requirements are as follow

Low: 0.01 - 0.2 amps AC

Moderate: 0.25 - 0.49 amps AC

High: 0.5+ amps AC

** Includes installation

*** 20% increase from typical lumber used on covered brdges

****Protection from Different Threat Levels

LV: Low Vandalism

MV: Medium Vandalism

HV: High Vandalism

LA: Low Arson

MA: Medium Arson

HA: High Arson

Figure 3. Equipment selection tool.

Natural Barriers: The bridge owner does not have much influence on the types of natural barriers present at a given bridge site. In most cases, these types of barriers cannot be moved or changed in any major way because of physical limitations, aesthetics, or environmental regulations. Thus, the security system designer should seek to leverage the natural barriers to enhance the security at the site and minimize any negative influence they may have on the system. Many different types of natural barriers will be unique to every bridge site. Examples of possible natural barriers that may be present at any given site are as follows:

- Rivers are an effective way of keeping trespassers away from the shielded and vulnerable underside of the bridge.
- Thick brush—Much like rivers, some covered bridges are in heavily forested areas where vegetation will be close to the bridge site. Some may consider this to be a nuisance, but thick brush can act as a natural barrier that will keep out motor vehicles and, in many cases, trespassers. The security system designer can choose to plant hedgerows in strategic areas to deny or deter trespassers from entering certain areas of the bridge including underneath the bridge itself.
- Mountains—Some bridges may be in mountainous areas where sheer rock walls or other features can be present around the bridge site. Sudden changes in elevation can aid the security system, as many trespassers would be deterred by having to climb a large shear wall or incline.

Man-Made Barriers: As mentioned earlier, several different options for barriers can be chosen by the security system designer. Some options may not be reasonable at all sites but some possible options for man-made barriers are as follows:

- Decorative planters—The use of decorative planters can take multiple shapes whether it is as a planter or a bench. The main advantage of using a decorative planter is that they are aesthetically pleasing and can be an efficient barrier. They will deter motor vehicles from entering the bridge site but can be temporarily moved by a forklift or other heavy machinery in the event that someone with authority needs to enter the bridge for maintenance or other legitimate reasons.
- Bollards—A bollard is a cylindrical tube that is usually 12 to 24 in. in diameter and can have variable heights above the ground (Fig. 4). The many types of bollards include fixed, removable, and retractable, and they can be constructed from a wide range of materials including wood, concrete, steel, or plastic.
- K-rail—Also commonly known as a Jersey barriers because they were first used on the New Jersey Turnpike, K-rails are typically used to deflect motor vehicles safely. K-rails are sometimes viewed as not very aesthetically pleasing (Fig. 5).
- Welded steel guard rails—Much like the K-rail, the use of welded steel guardrails may not be the preferred choice because of its negative aesthetic value. Although it may have not been aesthetically

pleasing, the use of steel guardrails is generally a low-cost and effective way to protect the entrance to bridges.

Figure 4. Typical bollards.

Figure 5. Standard K-rail (Permission to reprint given by Rob Vander Veen, Mid State Concrete Products, Santa Maria, CA).

- Berms/ditches—One of the simplest ways of creating a barrier is to simply have a ditch or berm around an area that is off limits (as long as it does not create a hydraulic scour hazard). This may not be as effective as other means, as some motor vehicles may be able to navigate over or around a ditch or berm.

Fences: In Chapter 3 of Perimeter Security, Michael Arata (2006) discusses fences in full detail in the book. Fences at a bridge site would help to force the public to enter the bridge site through secure entrances that can be monitored and controlled. Fences should not be considered to be a way of completely preventing unwanted entry into a bridge site but rather a way of slowing down and deterring trespassers.

When deciding to build a fence, many decisions need to be made, depending on the type of bridge site that is being secured, including the following:

- Chain Link—Chain link fences are the most common type of perimeter fences because of their price, easy set-up, and low maintenance requirements. The chain link fence is so functional that the Federal government widely uses it and has the following specifications that can be found in Federal Standard RR-F-191/1A. The following is a summary of Federal specifications taken from the Defense Logistics Agency (DLAI 5710.1):

1) Fabric made of chain link
2) No. 9 gauge or heavier wire
3) Seven feet high
4) Fence fabric mounted on metal posts set in concrete
5) Mesh openings not larger than 2 in^2
6) Fence bottom within 2 in. of solid ground
7) Fence top guard strung with barbed wire, and angled outward and upward at a 45-degree angle
8) The fabric used for the chain link fence should be galvanized, aluminized, or plastic-coated woven steel. The fabric should be connected to the posts with the same gauge wire that the fabric itself is made out of. If a fabric is used that has openings larger than 2 in^2, then the fence itself will be much easier for intruders to climb and should be avoided when possible. The use of privacy slats are commonly used with chain link fences but should be avoided on bridge sites as they allow intruders the ability to approach the fence without detection from the outside.

The Federal specifications are not required by bridge owners but are good guidelines for creating an effective perimeter. The use of barbed wire for extra security may not be required, as most bridge sites will not need this high level of security.

- Wrought Iron—The wrought iron fence is growing in popularity but is used mainly as an upgrade to chain link fences. Wrought iron fences are used commonly in residential areas as they are more decorative and aesthetically pleasing. The top of the fence is typically bent toward the outside of the area being contained with the ends sharpened for added security. It can be more difficult to climb a wrought iron fence than a chain link fence, especially a chain link fence without barbed wire, but the added cost of a wrought iron fence may be prohibitive.
- Wood—There are many options when using wood fences including design and level of security. Most wood fences do not allow a clear line of site to the other side of the fence, which is not desirable on a bridge site. With this lack of vision and the increase in maintenance required when compared with steel fences, wood fences are not typically a viable option for bridge sites.
- Concrete or Block Wall—Much like the wood fence, when using concrete or block walls many options can be considered as each wall is unique (e.g., type of block, color of block, type of foundation). Many concrete walls have barbed wire installed on top for added security, but this is not a requirement. Concrete or block walls will generally be significantly more expensive than a steel fence and will take longer to install.

Detection Equipment

Infrared Camera: An IR camera has the ability to detect and record the temperatures of different elements within its field of view during all times (e.g., during daytime and night time). This key attribute of IR cameras makes them especially attractive, as most acts of arson and vandalism occur during at night when the perpetrator can be masked by darkness. The major drawback for installing an IR camera at a bridge is the inherent cost when compared with other surveillance equipment.

IR cameras with appropriately configured software have the ability to detect motion and to issue alarms at certain temperature thresholds, and certain rates of temperature change. Great care should be taken when using a motion detection device of any type as large animals, such as deer or bears, could enter the field of view of the camera and set off the motion detector resulting in a false alarm. In general, however, these problems can be overcome by properly setting the "person" detection temperature threshold properly. Another threshold temperature setting issue relates to the type of wood preservative used on the bridge. Specifically, it has been observed that timber treated with creosote can get hot enough to approach flame temperatures and can thus cause false alarms.

As mentioned earlier, the main negative associated with IR cameras is the initial purchase price. Since IR camera technology is more advanced than other surveillance equipment, it might be more costly. As a result, the designer must consider if the advantages of deploying IR camera technology warrants the initial cost.

Typically, IR cameras are not built for exterior applications so it is important to include weather-resistant, protective housing for the cameras when estimating the system cost. Many such enclosures are offered by the IR camera manufacturer and can be custom made for any sort of application (including internal heating, internal cooling, etc.). Although the cost is high, it is still a valuable piece of equipment and was used on multiple case study bridges detailed later on in this report.

Video Surveillance—Video surveillance equipment is viable for monitoring bridge activity during the daytime or if the bridge is adequately lit during nighttime hours. It is important to know the technical abilities required of the camera and to choose a camera that will meet or exceed the expectations. The price for a common video surveillance camera is quite low and it may prove appropriate to place multiple cameras around a site.

Some cameras have the ability to change from infrared to visible light depending on the level of light in its vicinity. These cameras may be a more economical decision than buying them both separately. There are several manufacturers of cameras for adverse exterior conditions that range in ability and price. It is important to work with the manufacturer to select the appropriate camera for the bridge site in question. Much like the IR camera, the video surveillance camera must be located in a tamper resistant and environmentally appropriate housing.

Heat Detector—Several different types of heat detectors can be selected for bridge security. The two most common heat detectors commercially available are the fixed temperature and the rate of rise heat detectors. The fixed temperature heat detector is set to sound an alarm once a predetermined temperature has been reached. A rate of rise heat detector will set off any alarm once a certain rate of temperature gain has been reached. Both the predetermined temperature and rate of rise can be adjusted. In many ways, these settings are relatively the same as the IR camera thresholds mentioned earlier.

Note that the rate of rise heat detector may be more vulnerable to false alarms within the given environment. Specifically, covered bridges may be susceptible to sudden increases in temperatures, such as being warmed by the sun or a piece of machinery operating next to or on bridges that are still open to traffic. With these conditions being present on most covered bridges, it is important to place heat detectors out of direct sunlight. Some possible places to put these devices would be underneath the wooden deck or underneath the rafters on the top portion of the bridge.

Smoke Detector—A smoke detector may assist with fire protection coverage on a covered bridge; however, it may have a higher source of false alarms because the smoke detector will have to be placed within the covered bridge to be effective and will be vulnerable to dust and other debris coming in contact with it. In general, most smoke detectors are not designed to operate outdoors and special care must be given if a bridge owner wishes to deploy such a monitoring device.

Flame Detector—A flame detector may be one of the more important pieces of equipment in a set-up for security of covered bridges. The flame detector may be placed within the bridge interior where it will be able to view a large portion of the bridge. If necessary, multiple flame detectors may be used together in a system to optimize coverage of the bridge. Most flame detectors have a mechanism that significantly reduces false alarms

In general, flame detectors rely upon multiple measurements to determine if a fire has been set. First, and probably most obviously, flame detectors detect heat and are generally preprogrammed to know typical flame temperatures. Further, flame detectors typically have a sensing component that detects the unique flicker rate of flames. These redundant systems tend to reduce the number of false alarms that occur.

Fiber Optic Sensor—Fiber optic sensors may be used to measure rotation, acceleration, electric and magnetic field measurement, temperature, pressure, acoustics, vibration, linear and angular position, strain, humidity, viscosity, chemical measurement, and a host of other sensor applications not mentioned here. The fiber optic sensors typically used for bridge security measure changes in temperature along the length of the bridge. The main advantage of these sensors is that they have good bandwidth size and are lightweight, small, passive, resistant to electromagnetic interference, highly sensitive, and durable.

Many different types of fiber optic cables vary in price and it is important to choose a cable that can survive in the adverse exterior conditions where many of the bridges are located. The cost of fiber optic cables can be very cheap when compared with other equipment, and the cost continues to go down as the technology improves. However, the hardware required to read the fiber optic sensors tends to be quite expensive.

Defend Equipment

Fire Hydrant—One portion of fire protection that could easily be overlooked is how the fire will be extinguished. Fire trucks can store their own water and some even have the capability of pumping water from sources such as nearby rivers and lakes. However, in some cases the water source is too far below the elevation of the road to allow pumping. In this case, an additional pump at the water may be necessary to aid the fire truck in pumping the water up to the road elevation. It is important to know the limitations and abilities of the fire department that provides service to the covered bridge of interest.

If a municipal source exists near the structure, fire hydrants should be located on both sides of the bridge placed at a far enough standoff distance that heat from a bridge fire would not impede fire fighters hooking up hoses. If it is uneconomical to connect to the municipal water source because of the isolated condition of the covered bridge, it may be plausible to install a dry fire hydrant by connection to a water source nearby, as mentioned above.

Sprinkler System—If a fire of considerable size has been established before the fire department can arrive at the bridge, the center of the bridge may be inaccessible to fire personnel. This is a major problem for the structural integrity of the bridge, as this is where some of the largest forces exist. This problem can be solved by the installation of sprinkler systems within the center portion of the bridge or along the entire length of the bridge. It is essential that the sprinkler system selected be a dry pipe system so that freezing and bursting pipes do not become a problem during inclement weather. A dry pipe system has no water in the pipes until a sprinkler head is set off, at which point water enters the system and extinguishes the flames.

The first question the bridge owner would ask when discussing the possibility of installing a sprinkler system is the aesthetic effect on the bridge. As most sprinkler systems are relatively bulky and unattractive, it is important to place them in inconspicuous areas that are not seen by the public. The most economical and effective areas to place the sprinkler systems would be running underneath the bridge with the flow pointing upward toward the bridge deck and above the roofing members with the direction of the flow pointing down and to the sides. This will provide the greatest bridge coverage for most bridge applications.

Fire-Retardant Material—Achim Hering (2001) gives us insight on fire-retardant materials in *The Proof Is in the Fire.* One of the most common types of security measures taken by covered bridge owners is to increase the fire resistance of the bridge. This is due to the cost efficiency of the material when compared with the level of protection added to the bridge. This can come in the form of some sort of coating such as intumescent coating or using fire-retardant wood when rehabilitating or repairing a bridge. This falls under the category of structural strengthening and will help the bridge stay structurally sound during the course of a fire before it is able to be extinguished. Using fire-retardant materials should not be seen as a final solution for bridge security because they do not stop the initiation of a fire but slow the progress of a fire such that major damage can be avoided. The application of a fire-retardant material to structural elements that are essential to the survival to the bridge is one of the most desired actions with bridge security related to fires.

Many of the lighter bridge elements such as shingles, siding, and any thinner wooden framing should be considered when deciding what needs to be treated with fire-retardant materials. These thinner pieces can be ignited much easier than the larger structural members such as the bridge deck and primary truss members. These thinner bridge elements can be the fuel required to ignite the larger members and cause serious structural damage in the case of a long-burning fire. Although it may be more economical and efficient to make the thinner elements fire retardant, it is not uncommon to make the entire structure fire retardant for an added level of fire safety. The cost should be weighed against the advantage of such an extensive procedure.

Intumescent Coating—One of the more common fire-retardant materials used for covered bridges is intumescent coating. Intumescent coating is a thin layer of material very similar to paint in method of application as well as aesthetics and texture. It can come in a variety of colors including clear and semi-clear. There are multiple manufacturers of intumescent coatings and similar materials that can match the exact color scheme of any bridge so this coating can be applied to existing or new bridges.

The way intumescent coating works is by swelling when exposed to a certain level of heat that would be seen during the course of a fire. The chemically bound water in the coating absorbs heat, making it ideal for fireproofing applications. Intumescent coatings are used to keep the fire in the location of its origin instead of spreading. This type of coating will only become active after being exposed to heat. Although the coating of essential structural members is the most important, it is also beneficial to coat the entire bridge including siding, roofing, and all framing.

Although this is a very effective method of making a structure more fire resistant, it still has several drawbacks that must be considered for optimum protection. Much like any coating, intumescent coating will break down and become less effective over time and will need touch ups every few years and possibly an entirely new coat after multiple touch ups. These times are dependent on the conditions at the bridge site including humidity, direct sunlight, and temperature changes. Different manufacturers will have different materials that will be adaptable to different conditions so it is important to choose the correct coating for an exact bridge location.

Fire-Retardant Wood—Robert Durfee (2003) has an extensive section on fire-retardant wood in his article, "Vermont's Covered Bridges." Fire-retardant wood is created by pressure-treating the wood with a fire-retardant chemical. Multiple manufacturers of fire-retardant materials use different chemicals and have differing levels of fire resistance. Most fire-retardant woods are designed to not light on fire even with direct contact with a flame or the fire will not spread after it is initially lit.

A major drawback to using fire-retardant wood is the decrease in the structural strength of the member by up to 10% to 20% depending on the exact type of chemical used. This can become a major issue if fire-retardant wood is used on structural members such as the bridge deck, as the members may need to be larger since they will have decreased strength. Much like the intumescent coating, there is a decline in the effectiveness of fire-retardant wood that is accelerated from being in a harsh exterior location. This decline in effectiveness can be slowed substantially by painting the wood soon after installing it.

Pole Placement

Most security systems, especially those involving cameras, will be required to have a pole or pedestal of some kind for placement of the necessary security monitoring hardware. Below are some of the key factors to consider when selecting a pole type and location.

Aesthetics

The pole placement will ultimately be a decision directed by owner of the bridge and the security system designer. The most important factor influencing the final placement of the pole will be the aesthetic effect on the area surrounding the bridge while at the same time being placed so that adequate coverage is achieved. Because covered bridges can be a major source of tourism for smaller towns around the United States, it is important that the security systems installed do not draw attention away from the bridge and the surrounding landscape. Depending on the landscape and surroundings, this may be detrimental to the effectiveness of the security system because this usually requires that the pole be placed a large distance away from the bridge. An ideal situation at the bridge location would be a pre-existing pole such as a light pole or power line pole that, upon approval of the owner, could double as the mounting pole.

The type of material used for the pole is mostly an aesthetic decision. Several different types of materials can be used for the pole such as metal, concrete, or wood. To blend in with the surrounding area, we suggest that a survey be taken of what is around the possible pole location to see if other light or electrical poles are within sight. If other poles are in the surrounding area, it would be preferable to use the same type of material and pole height to let the security pole blend into its surroundings.

Line of Site—Even if a pole is within a reasonable distance from the bridge, it must be guaranteed that there is a direct line of sight to both the front and rear entrances to the bridge to allow the system to be as efficient as possible. Even if there is a direct line of sight at the moment that the security system is enabled, it is important to look for possible obstructions in the future such as trees or other vegetation that could have the potential to grow or move into the direct line of sight. Any potential problems should be dealt with as soon as possible to avoid costly problems in the future.

It may be possible to only allow a direct line of site to one entrance of the bridge because of environmental constraints. If the designer has the ability to use a pole facing toward either entrance of the bridge, many criteria must be considered. If the bridge is closed to vehicle traffic, then the entrance that will see the greatest amount of pedestrians would likely allow for the greatest amount of security on the bridge. The ease of connecting to electrical power should also be considered if electronic equipment will be used in the security system.

Functionality—Placing the pole at a significant distance from the bridge creates multiple problems, and there must be a compromise between aesthetics and functionality to make the pole as practical as possible while still keeping its presence as benign as possible. The two major problems with placing the pole at a significant distance from the bridge come from the effectiveness of the cameras and problems associated with running the wire over long distances. Both of these problems can be solved by using more sophisticated equipment; however, doing so directly correlates to an increased cost for the project.

A camera is still effective in generating a clear and usable image at differing ranges. These ranges will obviously change with differing manufacturers and cameras, and it is important to be cognizant of all camera specifications to ensure that all limitations at a certain bridge are overcome by the camera selected. This is applicable to all cameras used on the bridge including infrared or optical cameras.

The other problem that results from a greater pole distance is functionality losses within the wires that must be run from the pole to the bridge. Although this particular problem will not usually be a critical problem with the security system design, it needs to be considered before placing the system in the field to ensure that the proper power is reaching the systems. If this is done, the system will operate as expected and will result in fewer maintenance issues in future. If major problems with power loss occur across the wire between the bridge and the pole, several different solutions could be applicable, depending on the amount of power loss. One solution could be to use a wire with a higher conductivity so there would be less loss. Another solution could be to use a higher source of power that will overcompensate for the loss across the wire.

It may desirable to the owner to use the pole for other functions outside of security reasons. These other functions could include placing a street light on the top of the pole to provide lighting for the surrounding area. This could also be considered a security device, as well-lit areas are less likely to have trespassers that could potentially cause harm to the bridge. It is important that the street light does not cause the area to be over lit, which could cause light pollution and take away from the aesthetic value of the surrounding area.

Enclosure Selection

Depending on the functionality of the security system, varying amounts and types of equipment that because of their design, construction, and cost will need to be housed in a secure location that will also protect them from weather. Several methods can achieve this, and much like the selection of all the other aforementioned components, several factors should be considered in the selection of the appropriate enclosure.

Types of Enclosures

Standard Box—A metal box will prove to be efficient for many bridge sites, as many pieces of electrical equipment require small components for power and storage of data. If cameras are used as part of the integrated security system on the bridge site, a small box could easily fit on the pole itself where all the cameras are located. If a pole is not available, a box may be positioned on some sort of pedestal to allow the box to be safely lifted off the ground as to deter any water, insects, or animals from attempting to enter the box.

NEMA Enclosures: It is generally suggested that all enclosures meet a certain standard set by National Electrical Manufacturer's Association (NEMA), and for most applications with these systems the owner should require at least a NEMA 4, 4X, 6, 6P. The technical explanations of the exact NEMA standards are as follows:

- Type 4—Computer enclosures constructed for either indoor or outdoor use to provide a degree of protection to personnel against incidental contact with the enclosed equipment; to provide a degree of protection against falling dirt, rain, sleet, snow, windblown dust, splashing water, and hose-directed water; and that will be undamaged by the external formation of ice on the enclosure.
- Type 4X—Protection unit constructed for either indoor or outdoor use to provide a degree of protection to personnel against incidental contact with the enclosed equipment; to provide a degree of protection against falling dirt, rain, sleet, snow, windblown dust, splashing water, hose-directed water, and corrosion; and that will be undamaged by the external formation of ice on the enclosure.
- Type 6PC—Enclosures constructed for either indoor or outdoor use to provide a degree of protection to personnel against incidental contact with the enclosed equipment; to provide a degree of protection against falling dirt; to protect against hose-directed water and the entry of water during occasional temporary submersion at a limited depth; and that will be undamaged by the external formation of ice on the enclosure.
- Type 6P—Cabinet constructed for either indoor or outdoor use to provide a degree of protection to personnel against incidental contact with the enclosed equipment; to provide a degree of protection against falling dirt; to protect against hose-directed water and the entry of

water during prolonged submersion at a limited depth; and that will be undamaged by the external formation of ice on the enclosure.

- For most scenarios, a NEMA 4 enclosure will protect the electrical components against the elements as submersion under water will not applicable to most bridge sites and the added cost of higher levels of protection will not be required nor warranted. If water coming from multiple directions is a problem for a bridge, then a NEMA 4X enclosure should be considered

Structure—Some bridge locations will already have a small mechanical shed or other enclosed structure on site. This could prove to be advantageous for the security system designer for its ability to house the mechanical equipment that is required for running a security system. Depending on the size and scope of the security system to be installed, it may or may not be economically viable to construct a new structure on the bridge site to house the mechanical equipment when compared with a box that could be attached to a pole or pedestal as mentioned earlier.

Aesthetics: The selection of the enclosure to house all of the electrical components of the security system is comparable to the selection of the pole. The box must not have a large aesthetic impact on the surrounding area but must maintain functionality. Most box set ups that are available to the owner will be constructed of metal and will have a grey or metal color to them. It is possible, and in many applications preferred, to paint the box to match the surrounding area.

Functionality: Once the design team has decided upon the exact NEMA type of enclosure that will be used for the project, it is important to decide on the exact dimensions of the box. This can be accomplished by taking all the equipment that will be placed inside the box such as modems, computers, power strips, and any other sensitive equipment that needs to be secured in a weather-tight enclosure. Once the team has a list of all the equipment that needs to fit into the box, it is important to draw a schematic of how everything will fit. Once a rough layout is available, take into consideration how many outlets will be needed to power all the equipment and exactly what type of power strip will be used. Once all this information is available, the team will have a rough idea of the size of the box. It is important to order a box that is larger than exactly what is needed for minor equipment changes, cords, and future expansion.

It is important to understand that the heat that will be generated from the electrical devices within the box will be exaggerated by the temperature outside of the box in summer. The temperature within the box can easily become hot enough to overheat the electrical components and shut off the entire security system. Purchasing a box that has a fan or louvers to allow air circulation will add cost to the project but will allow the security system to operate through even the hottest months. After-market fans and louvers are available to install on boxes that do not have adequate ventilation. It is essential that any openings in the box protect the components inside from water or insects entering and potentially destroying the electrical components.

Power Considerations

The remote locations of many covered bridges within the United States often make it difficult or expensive to gain access to electricity to power the security system. The distance between the closest grid power and the location of the bridge may sometimes be so great that running a power line to the bridge is not cost effective. If grid power is not an economical option, different types and combinations of renewable energy sources may meet the needs of the system. Renewable energy will be discussed in full detail later on this section of the report.

Grid Power

If a hook up to grid power is readily available and economically viable, it will typically be the best choice as it will be the most dependable source of electricity. It is important to have a constant source of power with any security system to ensure that it works with the least amount of down time. The only problem with grid power is the reoccurring cost from the local electric company for using their services. Given that the draw from a typical security system is relatively small, the charges will also generally be minimal.

Renewable Energy

It is prudent to consider using multiple sources of renewable energy when setting up this type of system because any one source will not be consistent at different periods of time. Wind power only works when a substantial wind is occurring and solar power will only generate electricity during daylight hours, with energy output varying with the intensity of the sunlight. By using both systems, you will increase the chances that one of the sources will be generating enough electricity for the security system. For times when not enough power is being generated, batteries must be used to store electricity if grid power is not available. It is possible, however, that one source of power will suffice if the draw from an individual security system is relatively small and the battery bank is relatively large and can provide adequate reserve power.

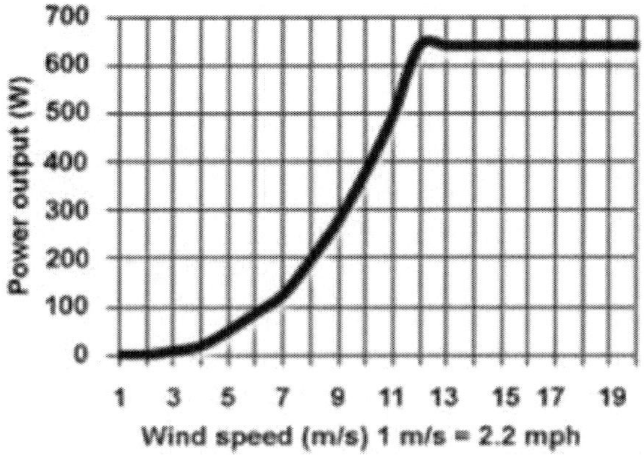

Source: Urban Green Energy 600-watt wind turbine specifications. Urban Green Energy Inc., New York.

Figure 6. Wind speed to power output graph for 600-watt wind turbine.

Using renewable energy to power a security system is a large undertaking and should be thoroughly considered before and during design. Compared with a direct city link for electricity, a renewable energy system will require notably more planning, maintenance, up-front cost, and patience. This fee will not occur with the use of renewable energy sources, but a city hook-up will not have the large up-front cost associated with setting up a renewable energy system.

Some systems may elect to use both grid power and a renewable energy system. A system of this type will use renewable energy whenever it is generated and use the grid as a backup if the renewable energy sources stop producing energy. By using this type of system, you can also use energy that is generated and stored. Some electric companies will allow you to generate electricity for them and will pay you incentives to do so. These incentives have the possibility of paying back the fee associated with using the grid power and possibly even pay for the security system itself.

Multiple types of renewable energy systems currently exist and several other sources are on the horizon. Although multiple sources of alternative energy exist, only a few will be useful at most covered bridges sites given their isolated locations and natural surroundings. The only sources of alternative energy that will discussed within this report are wind and solar power, as these are the two sources that will likely be present at most covered bridge sites.

Wind—Blades of a wind turbine harness energy from the wind and turn it into electricity through the movement of the blades. Ideally, a wind turbine should be placed in an area that has constant, nonturbulent wind. We recommend that the wind turbine be placed at least 30 ft above the ground and 300 ft away from all obstructions that could cause turbulent wind (e.g., structures, trees, changes in terrain). These distances are only recommendations and may be shortened, but optimal performance of your wind energy system will be at distances that equal or exceed these recommendations. Different companies may have different recommendations for placement of their particular wind turbines, and it is important to follow them as closely as possible to get the maximum efficiency from the equipment.

In general, the more power required from a wind turbine, the greater the blade diameter required. The blade diameter often becomes a limiting factor when designing a hybrid energy source depending on what is allowed by city ordinances and historic structure limitations. It is important to realize that the wind turbine will be placed within close proximity to a covered bridge that is a draw for tourism and the local economy, so care must be taken to not detract from the aesthetics of the surrounding area.

When sizing the wind turbine for a particular security system, it is important to realize that the rated output for the turbine is for the optimum wind speed, which will not be the average wind speed in many instances. These speeds can meet or exceed 20 to 30 miles per hour with most types of wind turbines, which is not a constant wind speed in most areas. It is important to review and consider the wind speedto-power output graph that is supplied with most turbines to choose the correct equipment that will produce the proper power output at average wind speeds at the bridge site. Figure 6 shows a wind speed to power output graph for a 600-watt wind turbine and shows that it will only produce 600 watts of power if the wind speed meets or exceeds 26.4 mph. This power curve has been measured under laminar wind conditions with the load connected directly to the generator.

For security systems with "lower" power requirements, 100 to 10,000 watts, there are two main types of wind turbines that should be able to produce the needed power: the horizontal axis wind turbine and the vertical axis wind turbine. Each type of wind turbine has its distinct advantages and disadvantages and there must be a benefit/cost analysis done to consider price, efficiency, and power output for the type of power output needed for a particular security system. The vertical axis turbines are well suited for smaller applications such as a security system and can survive wind gusts better than most horizontal wind turbines. It is important to discuss with the manufacturer the exact requirements needed for the security system and the environment where it will be installed.

Solar—Solar panels harness the sun's energy and efficiently convert it to electricity. Solar panels continue to become more efficient as the technology improves, making them smaller and more efficient; that said, size is still a main factor in the selection of a solar power system. Current technology allows 10 watts per square foot of solar panel according to top solar companies available for distribution in the United States (Anon 2011). Some panels will give you more or less wattage per square foot depending on price and the advancements in technology within the company. It is important to realize that an increase in efficiency will generally relate to an increase in price.

Although solar power can be an efficient source of renewable energy, it should not be thought of being without limitations. Because most security systems will need to be functional at all hours of the day, it is essential that a constant source of power is available day and night. Therefore, because solar power is only available during the day, it is essential to incorporate an adequately sized battery bank and potentially couple it with wind power or grid power to allow for uninterrupted power day and night.

Battery Bank—Batteries are important for any renewable power source system for moments when not enough sun or wind can generate the power needed to run the monitoring equipment. To design the proper battery bank for your system, it is important to know the power draw of your security system at peak performance, average hours of sunlight in your area, and the average wind speed in your area. Most renewable energy companies will be able to help you estimate the natural conditions at the bridge site depending on which part of the country the bridge is located. There are also multiple resources online with this information.

Batteries are designed specifically for renewable energy sources, but marine or deep cycle batteries are also quite effective if designed and sized properly. Because the system will create a steady draw from the batteries even though the batteries will not get a consistent charge, it must ensure that the batteries will not fall below 50% of their amp hour capacity so no damage occurs to the batteries. Once batteries have fallen below a certain percentage of their amp hours, they may not be able to become fully charged with the aid of a renewable energy system.

Inverter—The inverter converts the DC power created by the renewable energy sources into usable AC power. Two main types of inverters can be used for this application: the pure sine wave and a modified sine wave. A modified sine wave inverter is less expensive but does not produce the type of power required by sensitive equipment. Pure sine wave inverters are more expensive but are the appropriate choice for this application. Much like the solar panels and wind turbines, the inverter is not 100% efficient, so it will be required to increase the power supply to compensate for this loss across the inverter. It is important to research different companies that produce pure sine wave inverters and purchase one that meets or exceeds the requirements for the particular security system.

Controller—A renewable energy source will continue to charge the batteries indiscriminately if left unattended, which may overcharge the batteries and potentially damage them. Therefore, a charge controller is necessary to monitor the level of charge of the battery bank and manipulate the level of charge to the battery as needed. There are as many types of controllers as there are renewable power sources, so it is essential that a manufacturer's specification meet or exceed the requirements of the power system installed.

SET-UP

Several important steps are associated with set-up of the security system to ensure that it is aesthetically pleasing and functional. Although many of the major pieces of equipment have been discussed in great detail in previous chapters, it is important to have an understanding of all auxiliary equipment and components that are required to make the entire system functional. These materials include the type of conduit, wires, encasements, and the equipment to attach the box to the poll. More equipment may be required for a particular security system, depending on the location and type of system that will installed.

Conduit

To prolong the life of all wires and equipment, it is important to use conduit to run wires whenever possible. Several types of conduit could be used, including PVC, metal, or plastic. The design team should run an economic analysis deciding how vulnerable the wires are underground and how much each type of available conduit costs. It is important to choose a safe option that will adequately protect all the equipment, but it isn't necessary to always choose the strongest alternative. All conduit should meet or exceed all fire and electrical codes in the area of installation. The conduit should also be strong enough to protect against animals or insects penetrating the conduit and destroying the wiring.

Wiring

Wires come in a variety of sizes and types that range in price and functionality. These two main properties of wire are two of the biggest decisions when deciding on what type of wire should be used for any particular piece of equipment. Some equipment will have a minimum requirement for wire size to reduce resistance and improve the functionality of the piece of equipment itself. Table 2 demonstrates that as the gauge number of wire decreases, the diameter increases and the resistance decreases. A decrease in wire gauge will also lead to an increase in price per length, so it is not always practical to use the smallest gauge wire to increase functionality unless it is economically viable. If there is considerable distance between the equipment and from the power source, then a decrease in wire size may be required to reach an acceptable level of functionality.

Table 2. American Wire Gauge (AWG) chart for diameter, area, and resistance

AWG	Diameter (mm)	Cross sectional area (mm^2)[a]	Resistance (ohm/m)
0000	11.7	107.0	0.000161
000	10.4	85.0	0.000203
00	9.26	67.4	0.000256
0	8.25	53.5	0.000323
1	7.35	42.2	0.000407
2	6.54	33.6	0.000513
3	5.83	26.7	0.000647
4	5.19	21.2	0.000815
5	4.62	16.8	0.00103
6	4.11	13.3	0.00130
7	3.67	10.6	0.00163
8	3.26	8.35	0.00206
9	2.91	6.62	0.00260
10	2.59	5.27	0.00328
11	2.30	4.15	0.00413

Table 2. (Continued)

AWG	Diameter (mm)	Cross sectional area (mm^2)[a]	Resistance (ohm/m)
12	2.05	3.31	0.00521
13	1.83	2.63	0.00657
14	1.63	2.08	0.00829
15	1.45	1.65	0.0104
16	1.29	1.31	0.0132
17	1.15	1.04	0.0166
18	1.02	0.82	0.0210
19	0.91	0.6530	0.0264
20	0.81	0.5190	0.0333
21	0.72	0.4120	0.0420
22	0.64	0.3250	0.0530
23	0.57	0.2590	0.0668
24	0.51	0.2050	0.0842
25	0.45	0.1630	0.106
26	0.40	0.1280	0.134
27	0.36	0.1020	0.169
28	0.32	0.0804	0.213
29	0.29	0.0646	0.268
30	0.25	0.0503	0.339
31	0.23	0.0415	0.427
32	0.20	0.0314	0.538
33	0.18	0.0254	0.679
34	0.16	0.0201	0.856
35	0.14	0.0154	1.08
36	0.13	0.0133	1.36
37	0.11	0.0095	1.72
38	0.10	0.0078	2.16
39	0.09	0.0064	2.73
40	0.08	0.0050	3.44

[a] Resistance of copper at 20 °C.

Conduit can be easily used for running wire for certain applications such as underground or up a pole, but it may be difficult to use conduit within the bridge structure itself. All wire that is run without conduit should have a covering that can ensure proper protection. Within the bridge itself may be the possible threat of insects and small animals such as rats, birds, and squirrels; it is essential that all wiring be adequately covered to ensure that the wildlife will not be able to penetrate the coverings.

Attachment Devices

All equipment must be attached to either the bridge or other surrounding structures including poles or other structures. It is essential that all equipment is attached as securely as possible with a minimum of disturbance; this is especially true within the bridge structure itself. Many covered bridges have existed for over 100 years and it is critical that the security system being installed does not change the original design or construction of the covered bridge to preserve the historic significance of the bridge. If large alterations are completed on the bridge, it is possible that the bridge may not be eligible to be placed on the Register of Historic Places as mentioned earlier on in this report. Whenever screws or other attachment devices penetrate into the existing bridge, it is important to use the smallest size possible to minimize the permanent damage to the bridge.

Encasements

When ordering the equipment for a particular security system, it is important to notice the ability of the equipment to handle an outdoor atmosphere. Some types of equipment will be designated for indoor-only applications unless certain guidelines are met, such as the use of watertight enclosures. This is an often overlooked part of the security system but is important to the longevity of the system. Some proprietary systems will have casements for their products to protect them from exterior exposure while others may require the use of another system.

Some encasements may also be used in certain applications to increase aesthetic value of the equipment. These types of encasements may appear in the form as a wooden box around the equipment to give the appearance of a bird house, as seen in Figure 7. By using the same wood as the surrounding bridge, the flame detector shown in Figure 8 is not as aesthetically distracting as it is without the use of an encasement system. The simple act of painting the existing encasement of equipment to match its surroundings can be beneficial and will not detract from its aesthetic effect.

Supplementary Equipment

For any security system involving electronic components, it will be required to have supplementary equipment so that all components can operate efficiently. As mentioned earlier, it is important that someone knowledgeable about security systems, such as an IT professional, assists in the installation of all electronic equipment. Most electronic equipment will have software that may require manipulation to have the most efficient settings possible.

Communication Devices

It will be desirable for some information recorded at the bridge sites to be sent to a remote destination through the use of communication devices. The information recorded by the monitoring systems will prove to be useless if they cannot be seen by others at remote locations such as local fire departments or police stations. There are multiple ways of transferring this information to outside sources, but for more bridge sites, the use of directional antennas to access wireless connections will be the best option. Directional antennas allow for optimum signal in areas where cellular coverage may be low, which is typical in some isolated covered bridge locations.

Figure 7. Flame detector in covered bridge without encasement (top) and with encasement (bottom).

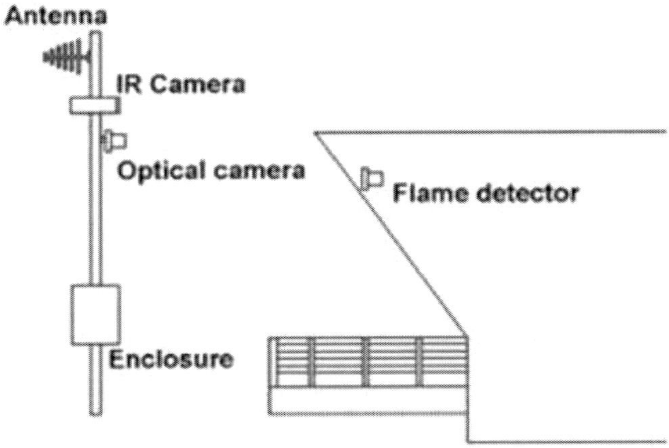

Figure 8. Cameras, antenna, and enclosure schematic at bridge site.

Storage Devices

It will be important that some data are stored at the bridge site if not all information is sent to a remote location. Depending on the abilities of the communication devices selected, it may not be possible to send all information required through wireless sources. A storage device can be as simple as a desktop computer or can be much more complex with secure electronic storage cabinets that can be found through multiple manufacturers. It will not be cost-effective in most applications to store all information that is recorded at the bridge site but to only record certain information. This can include certain time periods after an alarm has been set off or during certain times of the day.

Software

Numerous software options are available for designing a monitoring system on a covered bridge. Some of these software options may be proprietary depending on the piece of monitoring equipment that is selected by the design team. For the system to be as effective as possible, it is essential that the correct software is chosen and the correct settings are selected within the software. In-depth software analysis is not discussed in further detail within this report because of the numerous options available. It is important to work with the manufacturer of the monitoring equipment to allow for optimum efficiency from all equipment within the system.

TESTING

Testing is one of the most important aspects of creating an effective security system. All components should be able to fulfill or exceed their individual assignments, and the security system as a whole should be able to achieve a high level of security for the bridge as a whole. It should be stated that all fire testing, both in the lab and especially in the field, should be conducted in the safest possible fashion to ensure that there is no damage to any personnel, equipment, or property.

In-House Testing

It is crucial that all individual components of the security system are tested in-house before they are installed and tested at the bridge site. It is easier to troubleshoot equipment that is not working properly in an easily accessible, controlled atmosphere as compared with the bridge site, which may have inclement weather and equipment in hard-to-reach areas. Not only should the individual components be tested individually but the entire systems should be tested together to ensure that all parts of the system work together to reach the end goal of adequate protection of the bridge site.

In-Field Testing

As mentioned before, testing in the field should be conducted in the safest possible fashion. Before creating an open flame around, on, or within the covered bridge itself, it is essential that all proper officials are notified to avoid any problems with passersby from reporting the testing crew. Once all officials have been notified that fires will be started on the bridge site, it is essential that safety precautions such as fire extinguishers and buckets of water are on hand and close by in case a fire gets out of control.

Different equipment will have different thresholds at which they will be set off and trip an alarm. A typical fire for a flame detector may be a 1-ft by 1-ft^2 fire from a distance of 60 to 70 ft. This can easily be simulated by placing an adequately sized fire proof pan on a cart so it can be pushed around the bridge to ensure that all areas of the bridge are secure. This is essential to ensure that all cameras and detectors are pointing in the most efficient direction, and any adjustments can be made before the system is considered fully operational.

Infrared or other types of cameras may be motion detectors or be set to a certain temperature threshold. These should be tested at night when they would normally be operational and monitoring the bridge site. A fire may be started at one of the openings or inside the bridge within the line of sight of the cameras to ensure that the entrances of the bridges are secure from trespassers.

SUMMARY

Covered bridges are an important part of history of the United States and must be maintained for future generations. Unfortunately, these bridges are being destroyed at an alarming rate from arson, vandalism, and neglect. It is essential that covered bridge owners know the importance of maintaining the structural and aesthetic integrity of their bridges. As stated earlier in this report, The FHWA Blue Ribbon Panel Workshop on Bridge Security Assessments has decided on five different levels of security to construct an effective defense against unwanted activities. These levels of security include deter, deny, detect, defend, and strengthen structures. The five parts of the security plan proposed by the BRP are equally important if implemented correctly. This report has taken an in-depth look at the different security and monitoring equipment that can be used at covered bridge sites.

What These Levels of Bridge Security Hope to Accomplish

Bridge owners that implement the levels of security discussed in this report should not expect their bridges to be indestructible but must realize that this increased level for security will greatly improve the chances that the bridge will survive for generations to come. It is impossible to make a structure perfectly secure but it is the responsibility of the bridge owners to make the structure as secure as possible within the economic and aesthetic limits of the bridge site and the financial situation of the bridge owner. As mentioned numerous times throughout this report, it is essential that all levels of bridge security are covered.

Maintenance

The implementation of these security and monitoring devices should not be considered the end of any security project. Constant monitoring of all equipment and periodic maintenance of both the equipment and the bridge itself are vital to the survival of the system and the bridge. It may be required to renovate the bridge both aesthetically and structurally throughout its life and both of these issues are discussed briefly in the recommendations portion of this report and also in great detail in *Covered Bridge Manual* by Phillip Pierce. Within years or even months, a monitoring or security system can prove to be obsolete and may not be able to fulfill its initial requirements.

RECOMMENDATIONS

Protection against arson and vandalism alone will not ensure the safety and longevity of the Nation's covered bridges. Other preventative measures must be taken as often as possible so the bridges do not succumb to other natural or manmade disasters. There is extensive coverage on the types of preventative measures that should be taken with historic covered bridges in the Federal Highway Administration Covered Bridge Manual published in April 2005 in Chapter 17, "Preserving Existing Covered Bridges." Topics included in this report are controlling water runoff, roof and siding protection, foundation support, regular cleaning, and fire protection. The FHWA report goes into other preventative measures for covered bridges (Pierce and others 2005).

When an extensive search for security and monitoring systems on covered bridges through the United States was conducted, we found that the most-used type of protection for covered bridges was the use of flame-resistant materials. These types of products tend to have a low cost for how effective they can be and may only need maintenance every 5 to 10 years. Most covered bridge owners may not have large sources of funding and a large scale monitoring system may not be financially reasonable. Bridge owners may also have difficulty with the heavy amount of maintenance that may be required for some monitoring or security systems.

Education

Physical protection is not always enough when it comes to the preservation of our national historic landmarks. Education can be an important tool when it comes to preservation. Integrating covered bridges into the school system can teach the importance of these structures to younger students both in areas that have covered bridges and areas that do not. These educational benefits should include knowing from what era these bridges came, why they were built the way they were, and why they should stand for as long as possible as monuments of past.

Not only should the benefits and historical significance of these bridges be discussed in an educational setting, but the legal ramifications of damaging these bridges must also be addressed. Many people may not fully understand that most of these covered bridges are Federally protected, and damaging them can come with large fines and even jail time depending on the severity of the crime committed against them.

Phases of Monitoring or Security System

Four important phases of deploying any monitoring or security system must be thoroughly discussed before any portion of the project begins. For larger bridge owners, such as state DOTs or local governments, it may be possible for all portions of the monitoring or security system to be completed in house but this may not be possible for smaller bridge owners that do not have the capacity. To avoid any problems with the system and ensure that the system is as effective as possible, it is essential that all parties involved are aware of all arrangements and expectations.

Design

The design of the monitoring or security system can be very challenging and time consuming, depending on the level of security desired and the complexity of the system. If an intricate monitoring system is chosen, then software and hardware will be required and this in itself can become a very expensive and challenging problem. For small bridge owners who do not have the capacity of resources for this level of design, then it may be appropriate to outsource the design of the system.

Installation

Installation of any security or monitoring system heavily depends on the complexity and scope of the system. If the system is not very technical, then the installation could be completed by someone outside the security profession. If there are multiple pieces of equipment that must work in unison, then it may be in the best interest of the bridge owner to have a third party install all the equipment. In many cases, the company or individual that designs the system may be the same one that installs the system to guarantee that it is done correctly.

Monitoring

It does not matter how well any system is designed or installed if there is no one to monitor it and respond to any alarms. In most monitoring applications, the system must be monitored by a designated person who is able to react to any type of arson or vandalism. It is important to know what the procedure will be if any alarm is set off at the bridge site. If there are cameras at the bridge site, it may be possible to remotely view the bridge site after any alarm has been set off to see if it is a false-positive alarm.

Maintenance

The implementation of these security and monitoring devices should not be considered the end of any security project.

Periodic maintenance of both the equipment and the bridge itself are vital to the survival of the system and the bridge. It may be necessary to renovate the bridge both aesthetically and structurally throughout its life, and both of these issues are discussed briefly in the recommendations portion of this report and also in great detail in *Covered Bridge Manual* (Pierce and others 2005).

MADISON COUNTY PROJECT

The authors of this report have taken part in a security effort with Madison County in southwest central Iowa to protect five of the six covered bridges in the county. The bridges included within this effort are the Cutler–Donahoe, Hogback, Holliwell, Imes, and Roseman. The Cedar Bridge had a similar security system installed in 2005 by a team consisting of many of the same members. The Cedar Bridge project will be discussed in some detail with a full report on the entire project also available.

History of the Covered Bridges of Madison County

At one point, there were 19 covered bridges in Madison County but because of various reasons, all but five have been destroyed. Although it is unfortunately common for covered bridges to slowly disappear over the years, there has been a push in recent years to protect these historic landmarks. The bridges of Madison County have received heightened awareness because of their increased fame in part from the book, *The Bridges of Madison County*. This book, written in 1992 by Robert James Walker, was quickly turned into a major motion picture by the same name in 1995 starring Clint Eastwood and Meryl Streep (Walker 1992). In 1993, the book received more attention by being named "the book of the year" by Oprah Winfrey, which raised the status of the book and bridges to even a higher level.

Unfortunately, this fame did not protect the six remaining bridges at the time. The Cedar Bridge, the main bridge in *The Bridges of Madison County*, was completely destroyed in 2002 along with a house that was a major landmark in the book. The Hogback Bridge, which also appears in the book, was set on fire in 2003 but the fire was quickly extinguished by local passersby. The Cedar Bridge was completely rebuilt in 2003–2004 and the Hogback Bridge was completely renovated from its damages. These appalling actions caused the County of Madison to team with Iowa State University's Bridge Engineering Center and the USDA Forest Products Laboratory to install security measures on the Cedar Bridge to dissuade any further damage to the structure.

Because of the effectiveness of this project in 2005, it was decided in 2010 that surveillance equipment of a slightly different variety should be implemented on the other five bridges. The remainder of this case study deals exclusively with the project in 2010–2011 and the five bridges included.

Security Equipment Installed

The same security system was installed on all five bridges included in the 2010–2011 project. The system included a total of two cameras and two detectors: an infrared camera, an optical camera, and two flame detectors per bridge. In addition to the security equipment, numerous communication devices were installed within an enclosure that allowed for storage or information and remote connection to the surveillance equipment via cellular internet connection. The only major difference between systems was at the Hogback Bridge where alternative energy sources were used to power the surveillance system. This alternative energy system is discussed in detail in the Hogback Bridge section of this report.

When compared with the Cedar Bridge Security System installed in 2005, the main difference is that the Cedar Bridge was outfitted with fiber optic temperature sensors. Because of the problems faced by the design team during installation and testing of the fiber optic cables, we decided to not use fiber optic technology on the Cedar Bridge. Fiber optic cables can prove to be an effective means to protect a covered bridge but multiple problems can arise if not installed and operated correctly, as discussed earlier in the equipment portion of this report.

All equipment in the security system for this project was chosen by an information technology professional with experience within the field of security systems. Advanced software settings are not discussed in this report, as they are extensive. Also, information involved in programming all equipment to work properly within the system and all proprietary systems will have differing system set-ups. It is important to carefully read all instruction manuals and ensure that all proper settings are chosen while designing a surveillance system so that the optimum capabilities of the system can be obtained.

The overall system monitoring is completed with a local PC running custom-developed software with communications to all devices being made via a local area network with wireless and wired communication capabilities. Communications outside of the local area network are handled by wireless cellular radio. Three subsystems are used by the monitoring system: a Web camera, an IR camera, and UV/IR flame detectors. From the IR camera technology, the system will monitor output from the camera and detect a predetermined maximum range in which a heat signature will cause the system to activate the alert status. The UV/IR flame detector system uses signal conditioners that convert voltage into readable digital values. When the flame detector detects the presence of a flame, it completes an electoral circuit. The voltage from this circuit is read by the monitoring system and if it falls within the positive voltage range expected, an alert status will be activated.

When an alert status is activated, the monitoring system software activates the optical and infrared cameras to begin collecting buffered imagery from the camera for a specified time frame. For this project, it was decided to record the optical camera for roughly a minute and a half and the infrared camera for roughly 3 minutes. Unfortunately, this recorded video was not able to be sent wirelessly and has to be retrieved manually by the design team. Also, during this time, an email message is generated and sent by the wireless cellular radio network connection to identified recipients indicating an alert status has been reached.

Both active and historical images are viewable by anyone who has the system passwords. This was made possible with a cellular modem and a high-gain antenna.

Communication Devices

As mentioned in the previous section, multiple devices used on this project were for communication purposes so personnel could remotely access the surveillance system by an internet connection or to enhance the performance of the surveillance equipment. These items include a wireless router, personnel computer, cellular modem, Web-based remote power switch, signal conditioners, micro servers, power supplies, and power injectors. There are many proprietary systems available for these types of communication devices and the following list of equipment choices are only the selections made the by the BEC design team based on previous projects with similar applications. It is important for any individual design team to choose equipment that can be integrated. The team must feel comfortable using and configuring the equipment.

Wireless Router—An 11G Wireless Nano Router (ESR1221 EXT) was used to extend a wireless signal to the equipment being used on the bridge. This specific type of router has an upgradeable antenna for an increased wi-fi coverage area.

Personal Computer—The personal computer used on this project monitored the entire system in real time. It was also used to store images that are taken by all cameras during the case of an alarm being set off.

Cellular Radio—In addition to a directional antenna, we used a wireless cellular broadband modem so that all information can be accessed from remote locations.

Web-Based Remote Power Switch—A specific type of remote power switch has a timer function allowing a reboot or shut down times without any actions by in-field personnel. This power switch and its software was used to allow personnel to remotely control four power outlets at one time.

Power Injector for Optical Camera—The power injector used for this project was the POE – IPX-INJ-C. The injector delivers both data and electrical power to ethernet-enabled devices using a single ethernet cable. This eliminates the need to place the ethernet-enabled device, such as the optical camera, near an outlet and gives more freedom to the security system designer on the placement of the device.

Equipment for Flame Detector—The following equipment was used exclusively for the flame detectors.

Signal Conditioners—Two signal conditioners are used for every flame detector at all bridge sites to reduce false positives.

Microserver—A microserver was used on this project to connect the signal conditioners to the wireless connection.

Power Supply—A power supply was used to power both of the flame detectors, and another was used to power the relays in both of the flame detectors.

Cameras and Detectors

As mentioned earlier in this section, each bridge site has a total of two cameras and two detectors including an infrared camera, an optical camera, and two flame detectors. Figure 8 shows the placement of each piece of equipment that is outside of the box. The antenna, IR camera, optical camera, and enclosure are located on the pole roughly 100 to 150 feet away from the bridge, whereas the flame detectors are located within the bridge itself. The exact placement for each bridge site is seen later in the section when each bridge is discussed individually.

Flame Detector—Two flame detectors were installed on each of the five bridges in Madison County on either end of the bridge positioned to cover the largest portion of the interior of the bridge as possible. There were multiple problems when installing these flame detectors including cross bracing and other structural members interfering with the most ideal line of sight. The final placement of the flame detectors for each bridge is shown later.

The flame detector selected for this project senses radiant energy in the ultraviolet (UV), visible, and wide band infrared (IR) spectrums within a 120° of vision. The settings for the flame detectors were set at the 50- to 60-ft range, as most of the bridges are roughly 100 to 120 ft long from opening to opening.

To transfer the signal from the flame detectors to the data collection hardware where its conditioners were located, a trench was dug from the equipment pole to an entry point on the bridge. For most of the bridges, it was easiest to trench to the wooden approach span between the abutment and the bridge opening. From the underside of the approach span, the wire could be easily routed through the bridge to the flame detectors without being seen from the bridge deck. The Cutler–Donahoe Bridge had a newly paved road between the pole and the bridge, so directional boring had to be used in order to not disturb the road. The directional bore had minimal disturbance except for the entrance and exit points near the pole and the bridge and would be an option for bridge locations where open trenching is unwarranted or undesired.

Figure 9. Infrared camera image at Cutler–Donahoe Bridge.

Infrared Camera—For this project, only one infrared (IR) camera per bridge was positioned at a certain distance from one bridge entrance, usually within 50 to150 ft to the nearest bridge entrance. The infrared camera used on all five bridges has the ability to produce thermal images in real time at a resolution of 384 by 288 pixels. An enclosure provided by the manufacturer was used to protect the camera from the elements and individuals who would want to damage the camera with projectiles or through other measures. The enclosure for the IR camera had a glass lens that was coated in germanium for optimum thermal transmissivity.

Figure 10. Infrared camera image at Holliwell Bridge.

Figures 9 and 10 show images that the IR camera produces. Each image shows the capture date and time as well as the spectrum of temperatures within its field of view, including high and low temperatures. For each of these bridges a person stands within the opening of the bridge. The IR camera points out this area of increased temperature when compared with the surroundings with an arrow and a box with the temperature in it. If these maximum temperatures reach a certain threshhold, then then the IR camera can set off an alarm.

Figure 11. Optical camera image at Cutler–Donahoe.

Figure 12. Optical camera image at Holliwell.

Optical Camera—Directly next to the infrared camera was an optical camera. The optical camera provided a similar service as the infrared camera except, instead of providing images in the infrared range, they provided images in the visual spectrum. These images were also transmitted in real time over the internet.

The camera chosen for this project is designed for outdoor use so no enclosure was required before installation. This camera has the ability to pan and tilt through remote controls that could be controlled remotely.

Figures 11 and 12 show the image that the Panasonic optical camera produces. Compared with the IR camera, this image is fairly basic and is used for simple surveillance purposes. When an alarm is set through the IR camera of flame detector, the optical camera will record video for a specified amount of time and store it for future viewing so personnel can see what caused the alarm to trigger and to have video evidence of any individuals who may be on the bridge during times when the bridge is closed.

Other Equipment—Box

The equipment box storage used at each bridge was the same. A NEMA 4 box, which is the desired classification of box to be used for most applications as stated earlier in this report, which measures 24 in. by 24 in. by 8 in. (H × W × D), was used to store the monitoring equipment at the bridge. A rough sketch (Fig. 13) showed considerable room for expansion but after placing everything inside the box in the field, it proved to have very little extra room with the inclusion of all wires and power supply cords. This can be seen in Figure 14 with the actual in-field layout of the NEMA enclosure.

Because the NEMA enclosure was built to be hung on a flat surface, brackets had to be fabricated to secure the boxes to the monitoring equipment pole (Fig. 15).

These brackets were made with basic 1/4-in. steel bars that had drilled holes for lag screws and bolts. At the four points where the box was attached to the brackets, an eye bolt was used so aircraft wire could wrap around the pole to provide additional support against the box swaying in the wind.

Challenges with Each Bridge

Cutler–Donahoe

The biggest obstacle with the Cutler–Donahoe Covered Bridge was its location. The Cutler–Donahoe has been moved from its original location to a city park within the downtown area of Winterset, the county seat of Madison County. Because it was located within a city park, it was more difficult to get close to the bridge to complete any necessary construction or attachment of any cameras. A newly constructed road between the ideal pole location and the entrance of the bridge caused multiple problems with trenching for the flame detectors on the bridge as well as getting power from the bridge to the pole.

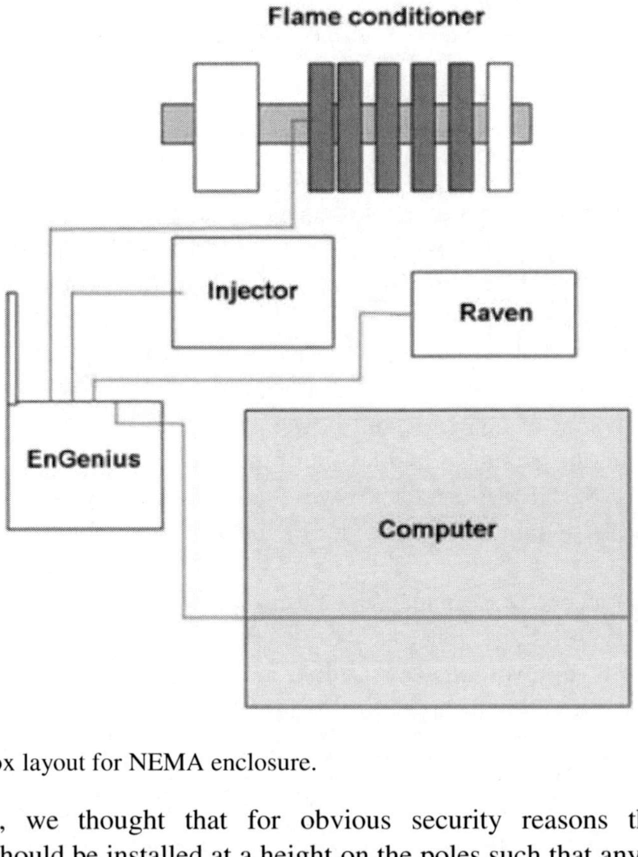

Figure 13. Box layout for NEMA enclosure.

Initially, we thought that for obvious security reasons the NEMA enclosures should be installed at a height on the poles such that any vandalism or damage would be difficult to accomplish. Therefore, for the first bridge instrumented, the Holliwell Bridge, the enclosure was installed approximately 20 ft from the base of the pole. Shortly thereafter, we determined this was not very user friendly, but also not necessary if other precautions were made. On the subsequently instrumented bridge, the Cutler–Donahoe Bridge, the NEMA enclosure was set roughly 8 ft off the ground as seen in Figure 16. We later determined that this height was still unnecessary and that setting the enclosures at chest height level was sufficient as long as enclosures were properly marked and securely locked. The bridge owner decided on the mounting height of all equipment. If the bridge site has had problems with vandalism, it may be necessary to keep all equipment at a higher level (8 ft or more) to deter trespassers.

Figure 14. Actual box layout in field.

The flame detectors for the Cutler–Donahoe Bridge (Fig. 17) were required to be placed in the middle of the openings because of the architecture of the bridge. This caused multiple problems because cross members tended to get in the way of a direct line of site for the detectors. Other bridges allowed side mounting of the flame detectors so that cross members and other structural members did not inhibit a direct line of a site and were placed optimally. It is essential when installing all equipment, both in the bridge and around the bridge site, to balance the aesthetics of the bridge site with the functionality of the equipment.

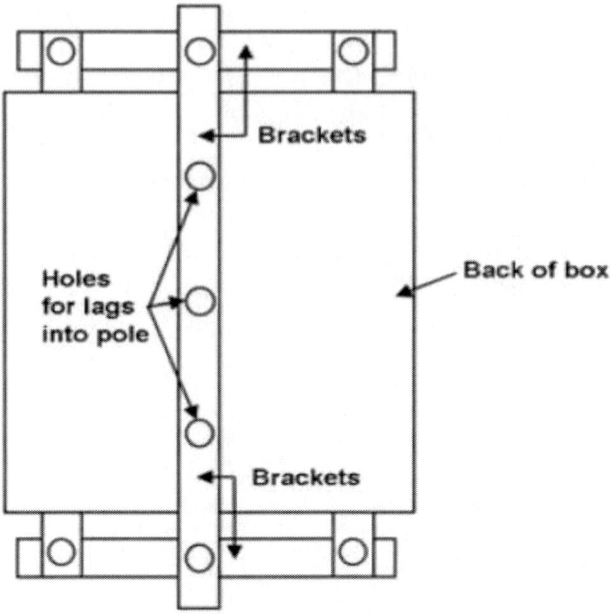

Figure 15. Brackets for box to pole connection.

Figure 16. Cutler–Donahoe pole.

Figure 17. Cutler–Donahoe flame detector that has been activated.

Figure 18. Hogback enclosure set up.

Hogback

The Hogback Covered Bridge proved to be the most difficult bridge out of the five instrumented. Because of the isolation of this particular bridge, it cost as much to install line power from the local municipality as it would be to set up a renewable energy system that used solar and wind power in the area. Neither the BEC nor the County of Madison had ever tried to develop such a large, stand-alone alternate energy system prior to this project, so there were initial uncertainties of such an undertaking. After discussing this, we decided to use alternative energy sources with the aid of renewable energy professionals. As seen in Figures 18 and 19, three additional NEMA enclosures for the renewable energy equipment and a 60-ft pole to house the solar panel and wind turbine were required. All of this equipment will be discussed in the following section.

Alternative Energy System—Because a solar panel or wind generator are only rated for ideal conditions, it was important to size all of the equipment for average conditions with a certain safety factor added in case of an extended downtime. After reviewing historical numbers from Madison County, we decided to select a 600-watt wind generator and a 150-watt solar panel with the use of a battery bank that could store reserve power for up to 3 days without stopping the security system (Fig. 20).

Wind Turbine—Instead of using a horizontal axis wind turbine, which is the more common type of wind generator, we decided to use a vertical axis wind turbine that could use less wind speed for more power. There is also a lesser chance that the vertical axis wind turbine will be damaged by sudden wind gusts when compared with its horizontal axis counterpart, which is ideal for the application in Madison County. Another difference between this system and others of similar size is that a common 60-ft telephone pole was used rather than a proprietary pole. Most manufacturers will only warrantee their equipment if it is placed on the pole that they also manufacturer, but some manufacturers will allow their systems to be placed on a professionally installed telephone pole. Significant savings can be seen by not having to purchase a proprietary pole for a wind turbine or solar panel.

Figure 21 shows the construction of the wind turbine within a laboratory setting for testing of power output at different wind speeds. For a frame of reference, the BEC employee in the yellow shirt in Figure 22 is approximately 6 ft tall. Figure 22 shows the complete assembly of the wind turbine in the laboratory setting.

Solar Panel

Battery Pack—Because batteries for solar panels are 12-volt and the power system is 24-volt, we were required to have four batteries in the configuration; two sets of batteries in a series and then the two pairs connected in parallel. Each individual battery had an estimated 225 amp hours and in the configuration of the battery bank was a total of 450 amp hours. Using only 50% of the amp hours available and with a total load of roughly 7 amp hours, the battery bank could provide power to the security system for up to 32 hours without charge from the wind and solar and with the security system running at 100% usage.

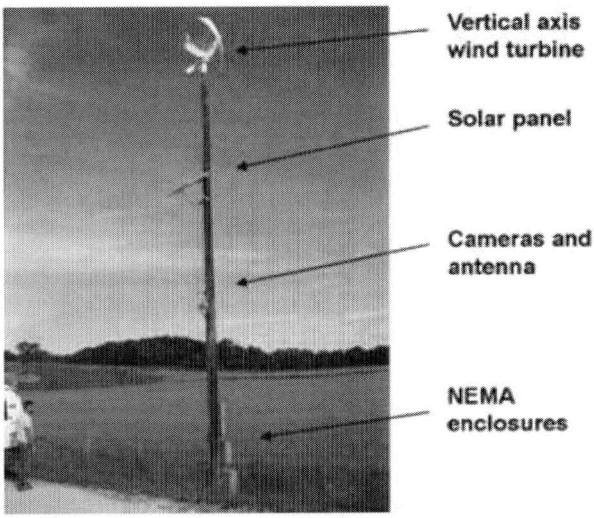

Figure 19. Original Hogback pole.

Controller—The controller used for this project was an off-grid controller designed specifically for the 600 vertical axis wind turbine (VAWT). The function of the controller is to take the variable 3-phase AC output from the permanent magnet generator and convert it to a stable 24-volt DC output to charge a battery bank. The controller also manages the turbine so that it performs safely and optimally.

Diversion Load—The diversion load is designed specifically to work with the controller for this particular set up. If too much charge comes from the wind and solar units at any given time, the excess power goes to the diversion load, which dissipates power through the form of heat. This piece of equipment is essential so that the system is not overcharged, potentially destroying the controller or batteries.

Inverter—The inverter used with this renewable energy system was a 24-volt DC-AC pure sine wave. As mentioned earlier in this report, it is essential that a pure sine wave inverter is used when powering scientific equipment. The efficiency of the inverter was 89%; therefore, there was an 11% loss between DC to AC. This loss was accounted for in sizing stages so that the loss did not affect the security system.

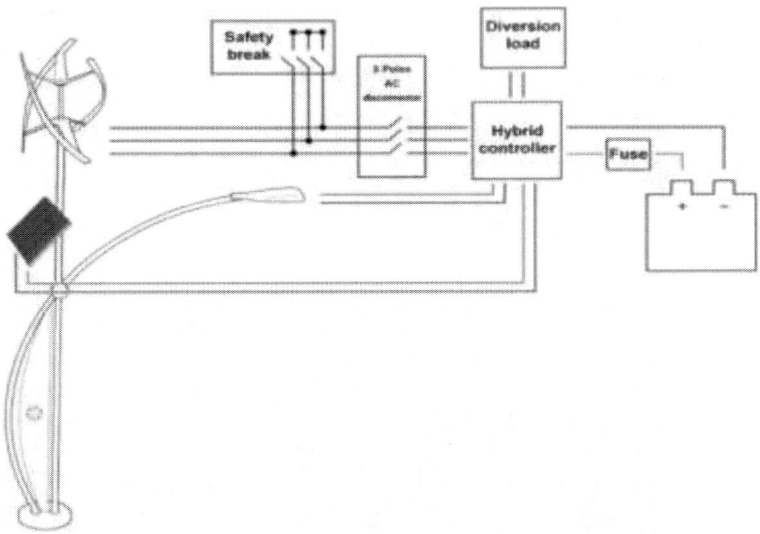

Figure 20. Wiring diagram for an off-grid system with solar and wind power.

Problems with Renewable Energy System

After installing the renewable energy, multiple problems arose with the power output of the system. The system could not support the monitoring system that had been installed for more than 4 to 5 days. We decided to slowly start expanding upon the renewable energy system while trying to decrease the power demand from the monitoring system by turning off certain pieces of equipment during different times of the day. The first change to the system was the addition of two 230-watt solar panels as well as a controller that was used solely for these two solar panels. Figure 23 shows the modified renewable energy system with the added solar panels. It can be seen from this photo that the extra solar panels are relatively large when compared with the original solar panel.

Figure 21. Assembly of 600-watt vertical axis wind turbine (VAWT).

Figure 22. Full assembly of 600-watt vertical axis wind turbine (VAWT).

Holliwell

Located between Winterset and St. Charles, the Holliwell Bridge was the first bridge instrumented. An instrumentation pole was placed on top of a significant hill with heavy weeds between the bridge and the pole. An above-head power line was the power source for the surveillance system. A pole that was roughly 100 yards away had power already established and we decided that using overhead power lines would be easier and more efficient than trenching between the two poles. As mentioned earlier, the heights of the NEMA enclosures vary between the different bridges and our initial thinking was to put the boxes high enough that they would be difficult for any vandals to tamper with but still be accessible with a ladder. After installing the box at the Hogback Bridge approximately 20 ft above ground, it was clear that this was not a safe situation for anyone doing maintenance on the system, and it was mostly likely not necessary if the boxes were appropriately labeled and securely locked. Figure 24 shows one of the optimum positions for the flame detectors to be placed within the bridge. By side mounting the detectors, there was minimum interference with structural members and the best line of site was obtained. Most of the mounting positions at other bridges were somewhere between the side of the bridge and the middle of the opening as with the Cutler–Donahoe and Imes Bridges. As mentioned in the Cutler–Donahoe section of this report earlier, it is essential that the aesthetics of the bridge site and the functionality of the equipment are balanced.

Imes

The Imes Bridge used an existing pole that was moved roughly 20 ft from its original position. The NEMA enclosure was placed at chest height, which proved to be the most efficient height for protection while still allowing easy access into the enclosure. The flame detector was placed much like the Cutler–Donahoe set up in the middle of the bridge opening. Existing power for street lamps was roughly 50 ft away and was hand-trenched to the pole.

Roseman

The Roseman Bridge was one of the easier bridge set-ups because an existing pole in an optimum position that already had power available. This allowed for minimum disturbance with the aesthetic value of the bridge site, as very little had to be changed to install the security system. The flame detector was positioned halfway between the middle of the opening and the side of the bridge. Because of the architecture of the bridge, this proved to be the optimum position for the flame detectors with direct line of site to the middle of the bridge.

Figure 23. Modified renewable energy system (Compared with original system in Figure 20).

Figure 24. Holliwell flame detector.

Figure 25. Required material for flame test.

TESTING

All five bridge sites were field tested to ensure that all equipment was operating properly. A 1-ft diameter pan was placed on a cart so a fire could be easily moved along the length of the bridge to ensure adequate fire-detection coverage. It should be noted that safety is the most important part of any in field testing that includes an open flame. Buckets of water and fire extinguishers should be close at hand as seen in Figure 25. It is also essential to inform all local authorities such as police and fire departments that open flame tests will be conducted to ensure any alarms or notifications from passersby can be ignored during a certain time frame.

The test started in the middle of the bridge span directly between the two flame detectors on either end. For some of the detectors to be set off, the fire had to be pushed roughly 10 ft toward the end of the bridge because the center of the bridge had some limitations to direct line of site.

The flame detectors were able to visually show that an alarm had been set off through the use of LED lights that could be seen through the wire mesh in the wooden boxes. A flame detector that is not set off and in working order is monitoring the bridge by blinking the LED lights every 15 seconds. Figure 26 shows a flame detector that was triggered during the fire testing. The red LED lights are constantly on and stay on until the system is reset via either internet connection or by simply cycling the power off and on.

After the flame detectors were found to be in working order, the same fire was started at the opening of the bridge on the side facing the IR and optical camera. Only the IR camera has the capability of setting off an alarm if a fire is present and there is no visual display to show if it is working properly as with the flame detectors. Once the IR camera would capture a temperature that exceeded the threshold temperature preset through the software, an email would be sent and the optical camera would capture a certain length of recorded video.

EXAMPLE CASE STUDIES

The creators of this manual decided to create multiple, fictitious case studies to show examples of how to implement different monitoring and security systems. These case studies show just one possible solution for each bridge. Numerous types of systems could be installed on each bridge depending on the requirements and financial resources available.

Figure 26. Flame detector after alarm activation.

Figure 27. Side view of Slaughter House Covered Bridge.

Slaughter House Covered Bridges

Bridge Owner
Town of Northfield
Bridge Location
Northfield, Vermont
Bridge history

The Slaughter House Covered Bridge is a 55-ft long queen-post truss bridge that carries traffic over Dog River by Slaughter House Road. This bridge was built in approximately 1872. This bridge is directly next to a highway and the local community is 1.5 miles away from the nearest fire department that has adequate equipment to put out a moderately sized fire on the bridge.

Figure 27 shows that the mid-span of the bridge is high above the Dog River below and virtually unreachable by local fire departments. This makes it more difficult to protect since a weakened mid-span can lead to total collapse. The lack of easy access deserves special consideration.

Security Option 1
Price: Low
Maintenance Required: Low-Moderate
Protection: Moderate
Equipment: Intumescent Coating, Signage

Because the covered bridge is close to a medium-sized city that has a fire department, the decision can be made to simply protect the bridge and reduce damage. By adding intumescent coating to the smaller bridge elements, such as the siding and roof structure, the bridge owner is drastically reducing the consequences of a fire. For added security, the bridge owner can decide to coat the structural members as well but because these members are harder to initially set, this may not be required. Adding signage to the bridge stating that the trespassers will be prosecuted will further increase the effectiveness of the security with little added cost.

Security Option 2
Price: Moderate
Maintenance Required: Moderate
Protection: Moderate-High
Equipment: Lighting, Flame Detectors

Adding a simple deterrent such as lighting can drastically reduce the chances that a trespasser will try to harm the bridge. This is especially true given the proximity to a major highway and a residential area. Because this option does not provide any protection, it is important that detection capabilities are high. This can be accomplished with the addition of flame detectors at one, or both, ends of the bridge. This will require higher maintenance with the addition of electronic equipment but will increase the security of the bridge.

Security Option 3
Price: High
Maintenance Required: Moderate
Protection: High
Equipment: Lighting, Sprinkler System

The most effective piece of equipment that could be used to protect this covered bridge is a dry pipe sprinkler system. Given the difficult accessibility, a sprinkler system could ultimately help the covered bridge survive if an arsonist was able to start a substantial fire before the fire department is alerted. Because this bridge is close to a residential area, we assume that a local water supply is close. Therefore, a hook up would not be as expensive as a more remote covered bridge. Adding lighting simply increases the protection of the security system by reducing the chance that a fire could be started in the first place.

Figure 28. Side view of Zacke Cox Covered Bridge.

Zacke Cox Covered Bridge

Bridge Location

Mecca, Indiana

Bridge history

The Zacke Cox Covered Bridge is a 72-ft single-span burr truss bridge built in 1908. It is in an isolated area with scattered farm houses in the vicinity. The nearest volunteer fire department is 4.9 miles away that has the capability of putting out a moderately sized fire.

Figure 28 shows a side view of the Zacke Cox Covered Bridge when the stream underneath it is completely dry. The stream is rarely completely full and can be easily walked across, allowing firefighters to get below the river and fight the fire from underneath the bridge if necessary.

Security Option 1

Price: Low

Maintenance Required: Moderate

Protection: Low-Moderate

Equipment: Intumescent Coating, Lighting

The simple addition of intumescent coating will reduce the chance that a major fire can occur at this bridge. The addition of intumescent coating on covered bridges is an attractive option for a bridge owner that does not want an expensive system or multiple pieces of electronic equipment. Maintenance will be required with this system but it will not usually require a completely new coat of intumescent coating for several years.

Security Option 2
Price: Moderate
Maintenance Required: Moderate
Protection: Moderate-High
Equipment: Intumescent Coating, Flame Detector

The addition of a flame detector will alert the fire department if a fire has been started on the bridge. This is important, as the fire department will have a 5-mile journey to get to the fire. Maintenance will be increased with the addition of a flame detector. However, this equipment will help to ensure that the fire department has as much notice as possible.

Security Option 3
Price: High
Maintenance Required: High
Protection: High
Equipment: Intumescent Coating, Flame Detector, IR Camera

The addition of an IR camera increases the level of protection from a fire by adding a level of security on the outside of the bridge. It can also be used to monitor individuals approaching the bridge at certain periods of the day. Given the remote location of the bridge, it is essential that all fires are detected as soon as possible. With the increase of electronic equipment, this will obviously increase the amount of maintenance required at the bridge site.

REAL CASE STUDIES

A national search was conducted to find different projects throughout the United States that included adding monitoring devices and different security equipment on covered bridge sites. Very few covered bridge owners were installing any monitoring devices, and a couple were adding other security systems such as lighting or fences. The majority of security installations dealt with structural strengthening and renovations to the bridges such as applying fire resistant materials to various components. These types of projects were not included in to the case studies as they did not add any sort of monitoring system on to the bridges.

Union County Covered Bridges

Bridge Names
Bigelow Bridge
Pottersburg Bridge
Culberson Bridge

Bridge Owner
Union County Commissioners

Bridge Location
Union County, Ohio

Initial Price of Security System
$215,619.00 (for all three bridges)

Bridge History

These bridges in Union County, Ohio, have had arson and vandalism problems but nothing major to this point. There have been fire problems from people starting camp fires on the river banks under the bridges but nothing that appeared to be intentionally set.

Security System Overview

The security systems on these bridges use a linear heat detection cable located along the edges of the truss space as well as one cable down the center of the ceiling space. A Fenwall system with a Honeywell remote dialer was used for the linear heat detection system. Two cables also run the width of the bridge above the abutment areas of each bridge (Cable placement shown in above image). Once the detection cable is set off by a certain temperature, an audible and visual alarm is set off and an alarm is sent to the fire department.

All four bridges were also equipped with Tokistar LED lights that were modified to withstand vibration from traffic on the bridges. The LED lights were located under the roof to aesthetically wash down the walls of each bridge. The LED lighting is used to make the structures more visible with the hope of reducing the chances of someone damaging the bridge during nighttime.

Knecht's Covered Bridge

Bridge Name
Knecht's Covered Bridge
Bridge Owner
Bucks County, Pennsylvania
Bridge Location
Springtown, Pennsylvania
Initial Price of Security System
$35,000.00
Source of Funding
Privately funded
Year Installed
2011
Bridge History
Bucks County has lost three covered bridges from arson since 1985 and almost lost another one when Knecht's Covered Bridge was set on fire twice over a 3-year period with the last attack in 2007. Hay bales were placed on the bridge with a trail of accelerant and lit by a match. Fortunately for Bucks County, the hay was damp, so it smoldered instead of becoming set ablaze.

Security System Overview
The security system for the Knecht's Covered Bridge included an alarm system with strobes and horns. These alarms are connected to a linear heat detection system that is controlled by a control panel enclosed in a NEMA 4 enclosure. There is also a dry standpipe sprinkler system across the length of the bridge to protect in case of any fire reaching a certain temperature threshold.

Pomeroy–Academia Covered Bridge

Bridge Name
Pomeroy–Academia Covered Bridge

Bridge Owner
Juanita County Historical Society
Bridge Location
Port Royal, Pennsylvania
Initial Price of Security System
$16,600.00
Bridge History

It is unknown exactly when the Academia Covered Bridge was originally built, but this bridge was destroyed by ice floating down the river in 1901 and was promptly replaced with a new bridge that still stands today. In 1962, the Pomeroy–Academia Covered Bridge was scheduled to be completely destroyed and replaced with a new concrete bridge to span the river. The Juanita County Historical Society took prompt action and acquired the bridge and allowed it to stand.

In June 2009, weeks after $1.4 million was allocated for restoration of the bridge, suspected locals attempted arson. The individuals lit articles of clothing on fire and damaged some of the bridge deck. A few days later, the same individuals, as thought by the local police, returned to extensively spray paint graffiti throughout the bridge.

Security System Overview

An eight-camera security system was placed throughout the bridge site as well as within the bridge itself for a total of $14,000 and all the graffiti was removed and a coating of fire retardant was applied to the burned area for a total of $2,600.

Cedar Covered Bridge

Bridge Name
Cedar Bridge
Bridge Owner
Madison County, Iowa
Bridge Location
Madison County, Iowa
Source of Funding
Federal

Bridge History

The Cedar Bridge was a major bridge in the 1992 book *The Bridges of Madison County*. In 2002, the Cedar Bridge was completely destroyed by arson. The town decided to rebuild the bridge with the same construction methods and materials as the original bridge and it was completed in 2004.

Security System Overview

An extensive security system was designed and installed on the Cedar Bridge with a grant from the USDA Forest Products Laboratory. This integrated system included an infrared camera, fiber optic temperature sensors, and two flame detectors at either end of the bridge.

Security System Technical Data

Major Equipment

One thermal imaging camera.

Two electro-optical digital fire detectors were used with the specification of the detection of a 1 ft^2 fire at 15 ft within 5 s. Twelve fiber optic sensors.

One optical sensing interrogator.

Additional Information

A detailed report for this project can be found in *Remote Monitoring of Historic Covered Bridges* (Phares and others 2010).

Illinois Covered Bridges

Bridge Name

Red Covered Bridge

Thompson Mill Covered Bridge

Bridge Owner

Illinois Department of Transportation

Bridge Location

Bureau County (Red Covered Bridge)

Shelby County (Thompson Mill Covered Bridge)

Initial Price of Security System

$74,450.00 (Red Covered Bridge)

$71,090.92 (Thompson Mill Covered Bridge)

Year Installed

2005 (Both bridges)

Bridge History

Before the improved security systems were installed on this bridge, some basic light fixtures were located in the interior and openings of the bridges. These basic fixtures included three wall-mounted luminaires located inside the bridges and one pole-mounted luminaire on the North approach.

Security System Overview

Two light fixtures were installed on the bridge during the security system improvement. These light fixtures will be controlled by photocells that turn the lights on when the ambient light levels fall below a certain threshold. A single architectural floodlight on the south ensures that both approaches are lit. A triple architectural floodlight is directed at the existing parking lot. After the renovation to the lighting system, the interior of the bridge, both approaches, and the parking lot are completely lit during all hours of the day.

Along with the increased lighting are five cameras placed throughout the bridge site. A camera in the entrance on both the north and south side of bridge faces toward the center so the entire interior of the bridge is under surveillance. A camera on the ends of both the north and south approaches faces toward the entrances of the bridge. A camera is also mounted with the triple architectural floodlight facing towards the mechanical building.

REFERENCES

Anon. 2011. "Solar & wind energy calculations: The (very) basics." www.solar-estimate.org. Energy Matters. Web. 21 June 2011.

Arata, Michael J. 2006. Perimeter security. New York: McGraw-Hill.

Becker, Fred J. 2011. History of covered bridges. Web. 24 May 2011. www.fredbecker.org/News Letter/ HistoryCoveredBr.htm

DOT. 2003. The Blue Ribbon Panel on Bridge and Tunnel Security. Recommendations for Bridge and Tunnel Security. Washington, D.C.: U.S. Dept. of Transportation, Federal Highway Administration, Print.

Durfee, R.H. 2010.Vermont's covered bridges, Vermont, tours, history, trusses. In: Covered Bridge News. 1 Aug. 2003. Web. 4 May 2010. Vermont Covered Bridge Society. www.vermontbridges.com/fire paper.htm

Hering, A. 2001. The proof is in the fire. Chemical Innovation. 31.5: 17-21.

Matias, E. 2011. National historic covered bridge preservation program - programs - integrated - bridge - FHWA.

Federal Highway Administration: Home. 21 June 2009. Web. 18 Feb. 2011. www.fhwa.dot.gov/bridge/covered.cfm

Murphy, K. 2008. Bridge arson part of crime spree. GoErie.com. 19 Dec. 2008. Web. 18 Feb. 2011. www.goerie.com/apps/pbcs.dll/article?AID=/20081219/ NEWS02/312199926/-1/NEWS02

National Fire Data Center. 2009. Intentionally set outdoor fires. Topical Fire Report Series. 9.6: 1-8.

Overington, C. 2003. Burn, Hollywood burn—bridges of Madison County threatened by arson. Sydney Morning Herald - Business & World News Australia. 1 Dec. 2003. Web. 18 Feb. 2011. www.smh.com.au/articles/2003/11/30/1070127272751.html?from=storyrhs

Phares, Brent M.; LaViolette, Michael D.; Wipf, Terry J.; Ritter, Michael A. 2010. Remote monitoring of historic covered timber bridges for the prevention of arson and vandalism General Technical Report FPL-GTR-191. Madison, WI: U.S. Department of Agriculture, Forest Service, Forest Products Laboratory. 27 p.

Pierce, P.C., Brungraber, R.L., Lichtenstein, A., Sabol, S., Morrell, J.J., Lebow, S.T. 2005. Covered Bridge Manual. Publication No. FHWA-HRT-04-098. April 2005. U.S. Department of Transportation, Federal Highway Administration, Research, Development and Technology Turner-Fairbank Highway Research Center, Mclean VA. Publication No. FHWA-HRT-04-098.

Tyska, L.A., L.J. Fennelly. 2000. Physical Security: 150 Things you should know. Boston: Butterworth-Heinemann.

Waller, R.J. 1992. The Bridges of Madison County. New York: Warner Books, Inc.

In: Historic Covered Bridge Security
Editor: Benjamin Donnelly
ISBN: 978-1-63117-041-6

Chapter 2

REMOTE MONITORING OF HISTORIC COVERED TIMBER BRIDGES FOR THE PREVENTION OF ARSON AND VANDALISM*

Brent M. Phares, Michael D. LaViolette, Terry J. Wipf and Michael A. Ritter

ABSTRACT

More than 700 historic covered timber bridges remain in the United States. In recent years, a number of these bridges have been targeted by vandals and arsonists. Obviously, prevention of future acts of vandalism and arson is important to retaining the invaluable parts of our engineering heritage.

A research team from Iowa State University, working cooperatively with the United States Department of Agriculture Forest Products Laboratory (FPL), developed a remote security and fire detection monitoring system for covered timber bridges. This system automatically notifies emergency response personnel in the event that a fire is detected or if unauthorized activity occurs. The system was installed on the Cedar Bridge in Madison County, Iowa in 2005.

Three independent detection systems were software-integrated and installed. Fiber optic sensors were installed to monitor ambient

* This is an edited, reformatted and augmented version of .S. Department of Agriculture, Forest Service, Forest Products Laboratory publication, General Technical Report FPL–GTR–191, dated February 2010.

temperatures throughout the bridge. An infrared camera, which is able to detect temperatures in an electronic image, was installed near the bridge. The data from these two parts of the monitoring system are used to detect rapid temperature changes due to fire and to also identify unauthorized persons on site. In addition, an industrial flame detector, which monitors radiant emissions within infrared and ultraviolet wavelengths, was mounted within the bridge interior. All of these components were connected using an onsite, wireless computer network with custom developed system integration software.

INTRODUCTION

Background

Covered bridges are unique structural systems. Typically designed and constructed in the mid- to early-1800s, covered bridges were originally devised as a way of constructing longer-lived bridges. By covering the primary structural components (heavy timber trusses and floor system) with less expensive and sacrificial coverings, bridge owners could extend the life of timber bridges. The principal need for this type of construction resulted because original timber bridges were constructed without modern preservatives. Therefore, when exposed to the environment, they tended to deteriorate relatively quickly.

Worldwide there are approximately 1,600 covered bridges, with more than half of these in the United States. Almost without exception, each of these covered bridges was designed for a specific location and to serve a very specific purpose in that unique location. As such, each bridge is essentially a one-of-a-kind work of art. Unfortunately, these irreplaceable structures have recently been vandalized and in some cases completely destroyed. Because of a best-selling novel and movie that featured the bridges, the covered bridges of Madison County in central Iowa are among the most famous of all the covered bridges. Unfortunately, one of these bridges, known as the Cedar Bridge, was completely destroyed by arson in the late 1990s.

Over the past 20 years, the Iowa State University (ISU) Bridge Engineering Center (BEC) has developed and deployed long-term monitoring solutions for bridges and other structures. Most recently, these activities have led to the development of unique monitoring systems that are autonomous and use state-of-the-art sensors. Coincidentally, the Forest Products Laboratory (FPL) has increased its interest in remote monitoring in housing and other applications. Because of the unique capabilities and the prospect for significant information and capability exchange, the FPL and the BEC began a collaborative research effort to address an important security need related to covered bridges. The results of that collaboration are the subject of this report.

Scope of Work

The objective of this work was to develop a cost-effective solution for providing security to covered timber bridges. The developed system was to be autonomous and provide security against what was believed to be the two most likely threats to covered timber bridges: arson and vandalism. Further, it was the intent that the security needs would need to be defined in two time regimes. First, fire detection would need to occur during all hours of the day with no reduction in sensitivity. Second, the detection of vandalism needed to occur only during select hours of the day, specifically, overnight. The two reasons for the latter restriction are that (1) all the bridges are open to the public and (2) because of less visibility, the most likely time for vandalism to occur was after dusk and before dawn. The developed system also needed to be capable of automatically and autonomously notifying local law enforcement and fire officials of the occurrence of both of the above.

Report Content

This paper describes several aspects of our research:

- Brief introduction to the topic of covered bridge security monitoring and description of the project.
- Review of published literature related to bridge security monitoring.
- Description of components and software that were developed and integrated to meet the goals of the project.
- Description of the testing on the developed security system. This testing was performed both in the laboratory located on the ISU campus as well as on-site following installation of the sensing components.
- Summary of the overall project and recommendations for future improvements to the developed system and for future research on this topic.

LITERATURE REVIEW

The following provides a very brief review of literature related to general security and bridge-specific security. In general, very little literature related to bridge security predates the terrorist attacks on September 11, 2001. Since that time, there has been new interest in identifying and securing bridges and tunnels that may be potential terrorist targets.

General Security

According to Stevens (2005), most security system design is completed using a multi-layered approach with "rings" of protection. When properly designed, these rings provide the best chances for detecting, evaluating, and responding to threats. Typically, each successive ring increases the level of security. Although there are variations in design, most include three rings of security that provide for (1) deterrence, (2) detection, and (3) delay. The two areas between the individual rings provide for locations or areas for incident response. The following generally characterize the three most common rings:

The first ring is generally designed to keep threats away. This can include such features as simple barriers to card readers and intercom systems. The goal of the first ring is to make a location unattractive to would-be threats.

The second ring generally tries to classify all persons within an area. In this instance, card readers and cameras that identify persons are typically used. In this ring, technologies such as revolving doors are also typically used.

The third and final ring generally tries to slow the progression of a threat to its intended target through the integration of electronically controlled locks and dual authentication devices. In general, these measures slow the movement of persons (both authorized and unauthorized) so that their presence and authority can be verified.

Bridge Security

As mentioned previously, little public consideration was given to bridge security prior to September 11, 2001. Following that day, a Blue Ribbon Panel on Bridge and Tunnel Security was requested to be established by the American Association of State Highway and Transportation Officials (Roberts and others 2003). That panel was asked to prepare a report on how to improve the security of bridges and tunnels and to develop strategies (both short- and long-term) and general guidance for improving security of major infrastructure assets. The resulting report documents much of the methodology that was followed in developing the conclusions. The framework by which security was to be addressed can be broken down into the following elements:

- **Identification of critical bridges through prioritization and risk assessment.** Although more detail is provided by the panel, the panel advocates broadly applicable and accessible prioritization methods and risk assessment based on rigorous engineering and mathematical principles.
- **Threats.** The following threats were considered by the panel:
  - Low- and high-tech conventional explosives
  - Penetrating devices
  - Low-tech, hand-held cutting devices
  - Truck and barge size explosives
  - Chemical and biological agents (tunnels only)
  - Incendiary explosives
  - HAZMAT release (tunnels only)
  - Intentional ramming via ship or barge
- **Damage.** The types of damage that are of concern include
  - Threats to integrity
  - Damage that inhibits functionality for 30 days or more
  - Contamination
  - Failure
- **Countermeasures.** Countermeasures were generally grouped into technologies that deter attack, deny access, detect presence, defend, or structurally harden. Because of the experience of the panel, only the last technology was given serious consideration. However, the panel noted that this consideration does not imply that the others are not valid options.
- **Knowledge and codes.** The panel basically concluded that, although some information is available, current codes and specifications are inadequate, and a significant research agenda is needed to provide the needed information.

From these elements, the panel developed seven primary recommendations categorized into three broad groups: institutional, fiscal, and technical. These are summarized below:

- Institutional
  - *Interagency coordination.* All stakeholders must collaborate to ensure that solutions meet the needs of all stakeholders.
  - *Outreach and communication strategies.* Information must be disseminated to decision-makers.
  - *Clarification of legal responsibility.* The Federal Highway Administration must seek to clarify the legal position of asset owners and owners must be advised of legal precedents.
- Fiscal
  - *New funding source for bridge/tunnel security.* Funds, beyond those already allocated by federal-aid highway sources, must be allocated from the Department of Homeland Security.
  - *Funding eligibility.* Federal funding guidelines must be amended to be independent of deficiency as it is currently defined.
- Technical
  - *Technical expertise.* The Federal Highway Administration and the Transportation Security Administration must collaborate to engineer all security solutions.
  - *Research, development, and implementation.* Research and development that leads to methods and standards must be a national priority.

From these seven recommendations, the panel developed strategies for improving security of both bridges and tunnels. These strategies are broken down into short-, mid-, and long-term strategies and are further sub-categorized as follows:

- Policy and planning
- Institutional continuity
- Review and prioritization
- Research and development
- Technology development and dissemination

In terms of providing general guidance, the panel made the recommendation that all security improvements should be considered in the context of mitigating threats and consequences. Basically, the guidance is "keep the bad guys out," and if they get in, "know in advance how to deal with them."

In January 2005, the American Association of State Transportation Authorities (Crossett and Rhodes 2005) summarized the role of State Departments of Transportation (DOT) in homeland security. In that document, the authors summarize, among other things, countermeasures in basic terms as the following:

- **Deterrence and detection** of attacks by securing access to structures or mechanical systems, improving lighting, conducting frequent patrols, and installing electronic detection systems;
- **Defense** against attacks by installing physical barriers that increase stand-off distances from vulnerable structural components, such as bridge piers or tunnel ventilation systems; and
- **Design and re-design** of assets to harden them against potential attack methods, particularly explosive charges.

Although several similar accounts exist, Cho (2005) reports on the installation of cameras and sensors by the New York Metropolitan Transportation Authority for the prevention of terrorist acts. Although for obvious reasons, many details are omitted, Cho recounts that the plan is to, "... integrate proven technologies such as closed-circuit TV, motion sensors, intelligent video surveillance and perimeter sensors into one seamless system..." and that the system, "...can be expanded to include additional technologies, such as biochemical detection and explosion detection." In light of the information presented above, this installation is an example of a context-specific system design that takes into account the likely threats and the need for future expansion capabilities.

Like some of the literature cited above, Williamson and Winget (2005) describe countermeasures to address threats to bridges as the following:

- **Planning and coordination measures.** These measures generally relate to the development, communication, and re-examination of plans to deal with threats.
- **Information control measures.** These measures are all related to sharing information among people and strive to give owners guidance on to whom and to what level information should be shared.
- **Site layout measures.** These measures (e.g., using lighting, vegetation, or landscaping) strive to improve the security of an asset simply by changing the environment in which it operates.

- **Access control/deterrent measures.** These measures include technological and policy changes that make it more difficult for threats to gain access to assets.

Deception measures. These measures complement those above by creating the appearance of a much larger web of protection. Because of the similarity in deployed technology, Wolff (2003) offers protocols for integrating bridge security into Traffic Management Centers. Although several significant barriers to this integration exist, Wolff offers the following general guidelines:

- Co-locate with law enforcement
- Co-locate with emergency management
- Determine critical bridges
- Explore methods to increase detection
- Explore methods to prevent access
- Establish alternate routes in the event that an attack occurs
- Assign a person or persons to monitor video of critical bridges
- Increase inspections and patrols of critical bridges during increased threat levels

SECURITY SYSTEM DESIGN

Initially, a combination of three sensing systems was proposed to detect vandalism and arson at relatively remote bridge sites. The first system consisted of fiber Bragg grating (FBG) sensors. FBG sensors, which are a specific type of fiber optic sensor (FOS), could be installed at strategic locations throughout the bridge so that they could monitor ambient temperature changes throughout and within the bridge. Second, an infrared (IR) camera could be installed in a location where it could view the entire bridge site and be configured to detect both the presence of humans during restricted times as well as the presence of a fire. The third system included a standard closed-circuit TV (CCTV) camera. Later, the addition of an industrial flame detector unit was added to the system following the discovery of an independent effort to install CCTV cameras at each Madison County covered bridge.

Bridge Site Layout

The Cedar Bridge, which was the bridge selected by the owner to install and demonstrate the developed system, is located approximately 2 miles north of Iowa Highway 92 near Winterset, Iowa (Figure 1).

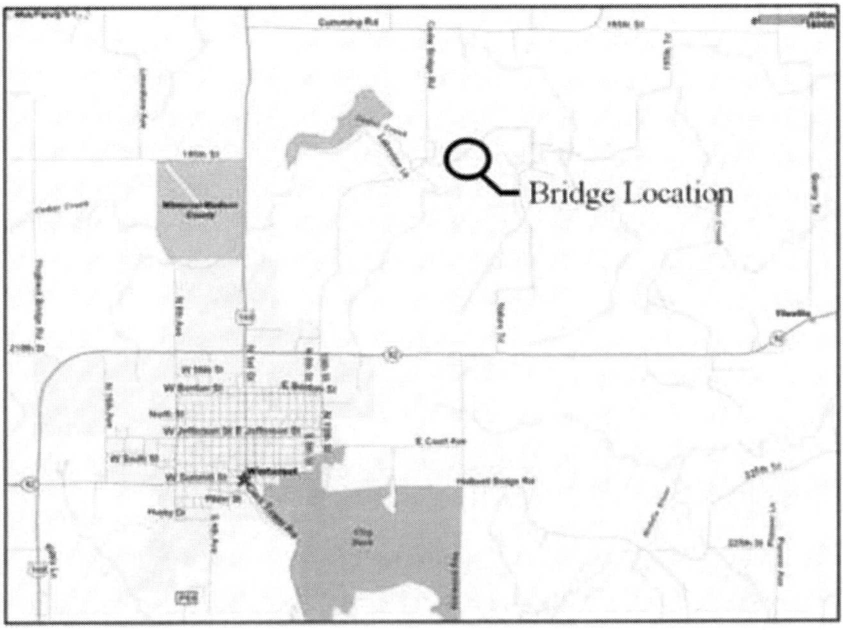

Figure 1. Cedar Bridge location near Winterset, Iowa.

The bridge is situated in a county-maintained park in a somewhat remote area with limited residences nearby from which the bridge can be observed.

The bridge crosses Cedar Creek approximately 300 feet from the nearest frequently traveled road. Further complicating the preservation of the bridge is that the bridge site is screened from the view of passing traffic by a large number of mature trees (Fig. 2). This type of remote location is common for the majority of covered bridges in the United States and is one of the reasons that remote security systems are vitally important to the protection of these historic structures.

Individual Sensor Components

Based upon a significant paper review of remote monitoring technologies, the research team decided that the security system would consist of three somewhat redundant sensing systems. The three sensing systems include infrared camera technology, commercial flame detector technology, and FBG sensor technology. The three individual sensing systems were to be integrated into a single system that provided overall detection.

Infrared Camera

A detailed examination was performed to determine the capabilities and limitations of infrared camera technology. To adequately monitor both the interior and exterior of the bridge, we originally thought that at least two infrared cameras would be necessary. However, upon the basis of further design and site limitations, we decided that a single, well-placed infrared camera could provide adequate protection.

Following a search for available IR camera systems, we contacted several manufacturers and supplied them with a site plan and general requirements for this project. The manufacturers were asked to recommend and provide a price quote for specific products that satisfied the project requirements. The recommendations received from the various manufacturers were as follows:

- Company A recommended a laser infrared illuminator at an estimated cost of $13,000. The illuminator would energize the surrounding area with infrared light and allow regular monochromatic cameras to "see" the infrared spectrum.
- Company B recommended two of their cameras. These cameras measure the temperature at every pixel in the field of view and cost from $10,000 to $20,000 each, depending on field of view and programming requirements.
- Company C recommended a 3-camera system (with proprietary software) with an estimated cost of $140,000.
- Company D recommended their camera, which produces infrared images but does not permit the user to measure specific temperatures within the field of view. The estimated cost of these cameras was $8,000–$9,000 each.
- Company E recommended their camera, which has a regular visual camera integrated with an infrared camera and is provided with a

customizable software package. The camera has an estimated cost of $50,000– $60,000.

Consideration was narrowed to two manufacturers based on the following criteria for this project:

- To detect the presence of humans during the hours of darkness, the camera must be able to measure specific temperatures within the JR range.
- The price of the JR camera must fit within the overall project budget.

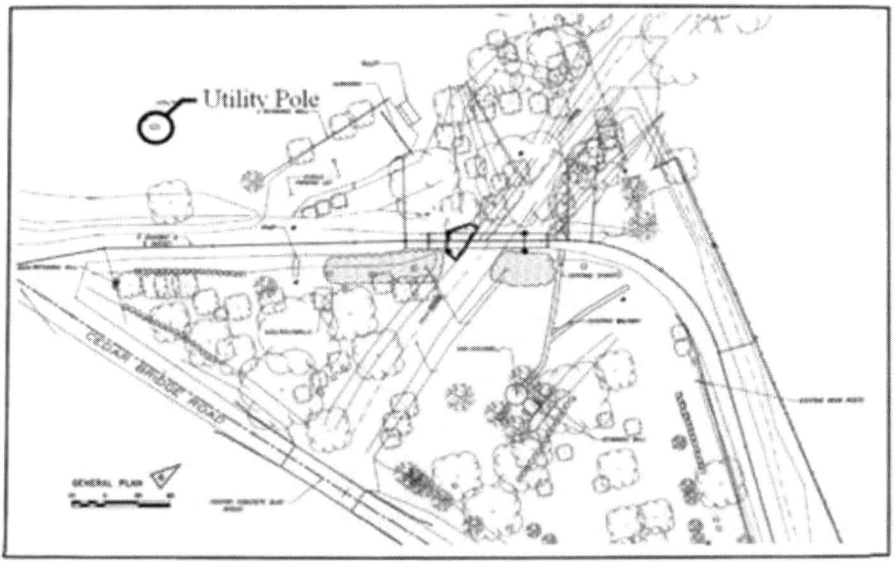

Figure 2. Bridge site plan view.

The cost estimate for the Company C system exceeded the entire project budget; thus, this unit was eliminated from consideration. The Company D camera and Company A illuminator did not permit specific temperature readings within an image and would not have been adequate for the detection of a fire event and were eliminated from further consideration.

The two remaining companies were asked to demonstrate the capabilities of their respective cameras. Company E demonstrated their camera (Fig. 3) using an online Web link and Company B brought their camera (Fig. 4) to the

BEC and provided indoor and outdoor demonstrations of the camera, including a summary of the software available with the camera.

J-rule.net, a small internet service provider in Winterset, Jowa, working independently of the research team, received a grant to install Web-based video cameras at all five covered bridge sites in Madison County. These cameras were also installed and maintained by J-rule. The BEC/FPL research team negotiated an agreement to work in cooperation with J-rule.net on the security monitoring of the Cedar Bridge.

Figure 3. Infrared camera.

The original research plan included the purchase and installation of CCTV cameras to observe the interior and exterior of the bridge. The high cost of the cameras, combined with the existing plans to install and maintain a Web-based CCTV camera at the site, allowed the research team to eliminate this part of the overall security system.

The Company E camera was eliminated from future consideration because the integrated visual spectrum video camera would no longer be needed. In addition, the proprietary software package supplied with this unit would not be able to interface with the custom software needed to control the various other sensing systems to be installed elsewhere on the bridge.

The selected infrared camera offered a number of advantages for the Cedar Bridge project:

- Optional software package
- Interchangeable lens with varying fields of view (12, 25, or 45 degrees from the axis)
- Lower price than the comparable, but newer, model camera
- Pole-mountable weather-proof camera enclosure included

Figure 4. Infrared camera.

The camera was purchased for approximately $15,000 and came with a pole-mountable camera enclosure, specifically designed for use with infrared cameras. The infrared camera was mounted on an existing utility pole approximately 300 feet southwest of the bridge (Fig. 2) to monitor the exterior of the bridge and the approach roadway. However, from this location, the camera is not able to monitor activities within the covered portion of the bridge.

Flame Detector

Recommendations from the ISU Department of Environmental Health and Safety (EHS) were used to quickly eliminate possible alternative fire detection systems for the interior of the Cedar Bridge. The alternatives that were considered and rejected included the following:

- **Standard smoke detectors**. They are considered unreliable in this environment, as road dust would likely result in numerous false alarms. **Electronic temperature gauges**. These would function quite well in this application, but would need to be located at numerous locations throughout the bridge and would essentially duplicate the function of the fiber optic sensors.

Based on recommendations from EHS, we searched for industrial fire detection sensors that would function reliably in a semi-harsh environment. In particular, focus was placed on flame detector devices that are designed to detect a fire based on the characteristic signature of a burning flame in three wavelength spectra (infrared, ultraviolet, and visible light) quickly and while the fire is still very small. The use of multiple light spectra sensors within a single detector unit, as it turns out, greatly reduces the occurrence of false alarms. These types of flame detector units are widely used in industrial applications (factories, warehouses) around the world.

In fall 2004, two industrial flame detector manufacturers were contacted. These two manufacturers offer very similar products, although one model was slightly less expensive. We ultimately purchased an ultraviolet–infrared electrooptical digital flame detector (Fig. 5). The specifications for this flame detector state that a 1 ft^2 fire can be detected at a distance of 15 ft within 5 seconds.

The flame detector outputs a fault circuit and a fire circuit. Each of these circuits operates in a binary mode and the voltage output by each circuit can be read using a multimeter or by other means. The fault circuit is tripped after the flame detector performs a self-check and finds a problem with its internal operation. The fire circuit is tripped when the unit detects the presence of a fire. To reduce the number of false detections, the unit was programmed to trip the fire circuit only when two of the three wavelength spectra independently identified a positive condition. The output combined with internal redundancy made integration of this sensing system relatively easy while also being very reliable.

Fiber Optic Sensors

The BEC has used FOS in numerous other projects to monitor changes in mechanical strain in structural bridge components (instead of traditional electrical resistance strain gauges). However, these sensors had not previously been used for measuring ambient temperature at a series of points throughout a structure.

A FBG sensor, a specific type of FOS, consists of a series of etched perturbations in the index of refraction along a short length of glass fiber (Figs. 6 and 7). This grating reflects a spectral peak and permits the remainder of the light to be transmitted along the fiber. The reflected spectral peak is based on the grating spacing and, therefore, changes in the length of the fiber (strain) will change the grating spacing and thus the wavelength of light that is reflected back to the source.

Figure 5. Flame detector.

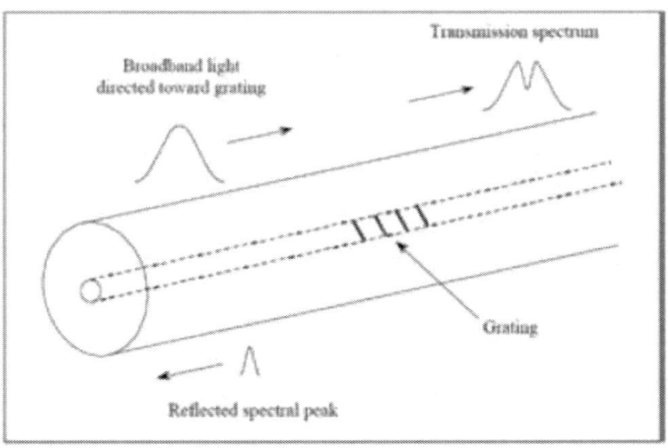

Figure 6. Fiber Bragg grating operation.

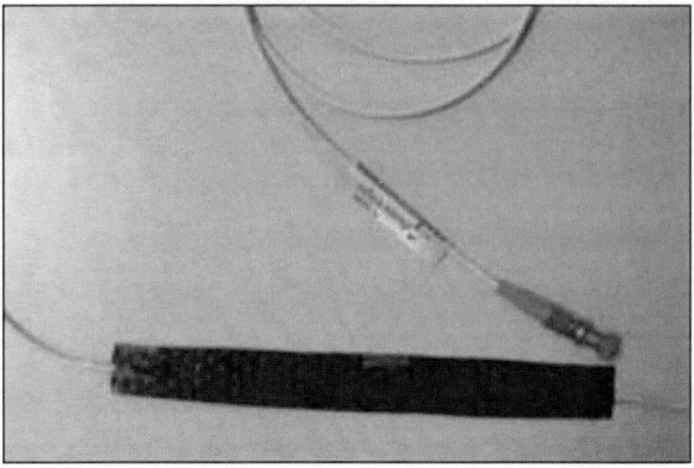

Figure 7. Fiber optic fiber Bragg grating sensor.

The center wavelength of each sensor in a series is spaced such that no signal overlaps between adjacent sensors. The leading end of the chain is connected to a device known as an interrogator, which transmits a pulse of very pure light along the glass fiber to each sensor. The reflected center wavelength of light from each sensor is recorded and compared with the initial center wavelength, and any change in wavelength is proportional to a change at the sensor location. Based on the change in wavelength, either a mechanical strain or temperature change can be calculated depending upon the application.

Prior to purchase of a fiber optic sensor system for the Cedar Bridge, a number of laboratory evaluations were made using an interrogator and sensors on loan from the manufacturer. These evaluations will be presented in the System Evaluation section and were used to determine how many sensors would be required to adequately monitor the bridge deck of the Cedar Bridge.

Security System Integration

The complete remote security system was developed using an onsite computer to collect data from the three sensing systems to determine whether a fire or intruder event had occurred and notify interested parties. The development of integrated software and hardware communications to perform this are presented in the following paragraphs.

Hardware Communication

An onsite wireless Local Area Network (LAN) was created using network interfaces supporting both Ethernet and wireless communication technologies. The onsite network consists of four devices:

- Flame detector
- Infrared camera
- Optical sensing interrogator
- Onsite desktop personal computer

The onsite computer was used to run the custom software developed to control the monitoring system. The computer system had a 700 MHz processor, 256 MB RAM, 20GB hard drive, Microsoft Windows XP Professional operating system (Microsoft Corporation, Redmond, WA), and a 10/100 BT Ethernet card.

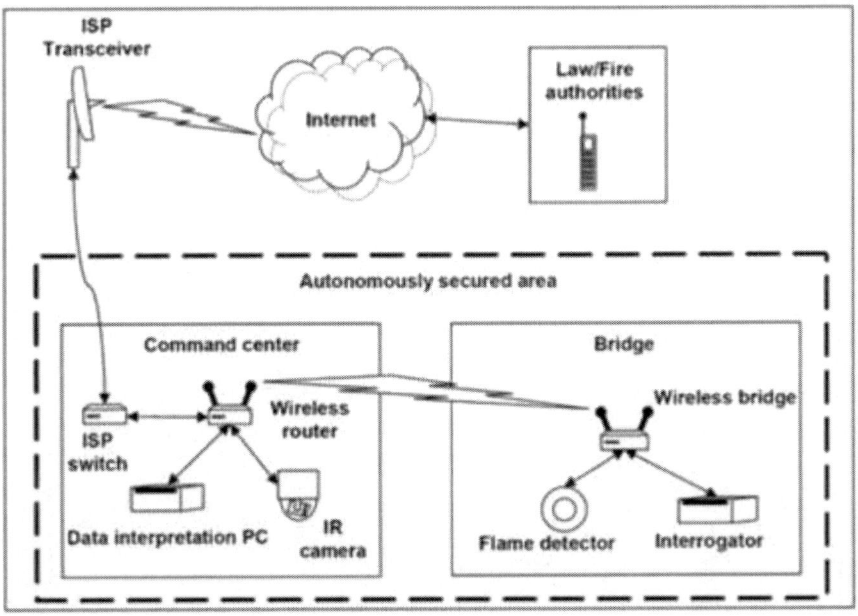

Figure 8. Schematic view of security system layout.

The camera and the PC were connected to a wireless router, and the flame detector and interrogator were connected to a wireless bridge. The wireless router and the wireless bridge were purchased at a total cost of approximately $150.

The flame detector required additional equipment to establish an ethernet connection to the system. A pair of signal conditioners connected to the fire and fault relay wiring of the flame detector was added to convert an output voltage from the device to a numeric output which could be read by a microserver attached to the wireless bridge through an ethernet connection. A 5-volt power supply was connected across the two flame detector relays, which were then connected to the signal conditioners. In this way, a zero-voltage reading indicates no fire or fault and a 5-volt reading indicates a fire or fault has been detected. The total cost of these components was approximately $750.

A schematic view of the onsite monitoring components is presented in Figure 8.

Software Communication

The software that monitors and controls all the sensing systems was developed using object-oriented ActiveX dynamic-link libraries (DLL) language modules. The DLLs contain reusable functions and data, and ActiveX technology, based on a Microsoft standard that defines how software components interact with each other. In total, researchers at Iowa State University developed four DLL modules—three to control each of the three major sensing systems and a fourth to control and integrate the other three modules.

The modules for each sensing system were written separately to decrease system maintenance time. In this way, when an update is required for one module, the entire program does not need to be recompiled—only the module needing to be updated. The main module, which controls the three component modules, is used to enable or disable the components that control any of the sensing systems.

Three possible events (fire, person, or fault) can be detected by the sensing system at the Cedar Bridge. Of these, the most significant is related to a fire, which may be detected at any hour of the day, or a person, who may be detected during late night hours. Each of the three sensing system modules was developed to send a true/false status report to the main module. Depending on the module, this status report represents either a "fire–no fire" or "person detected–no person detected" or a "fault–no fault" report.

The function of the four DLL modules is as follows:

- Main module—manages the inputs and outputs from the other modules. The main module only takes action when a fire, fault, or person event status is reported by one or more of the sensing systems.
- Flame detector module—sends two types of status reports to the main module: fire or no fire, and fault or no fault.
- Interrogator module—sends only one type of event status report to the main module: fire or no fire.
- Infrared module—sends two types of event status reports to the main module: fire or no fire, and person or no person.

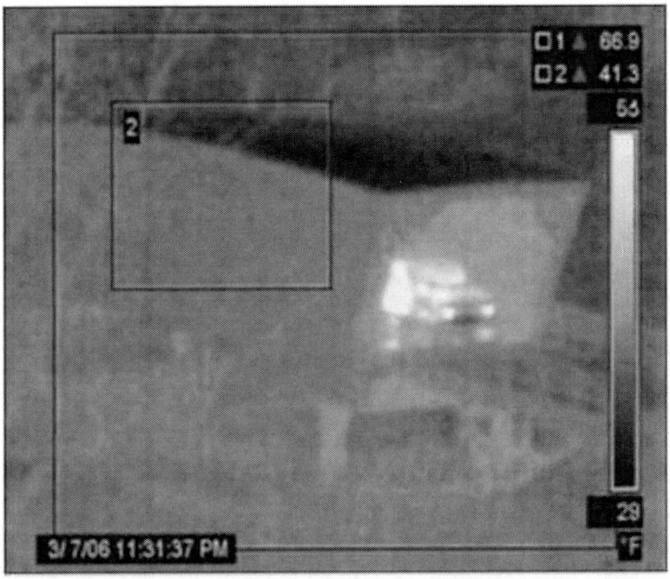

Figure 9. Sample image taken from onsite infrared camera following a person event.

If a fire event report is received by the main module from any of the sensor modules, the following events occur:

- The main module sends this event status to the other two sensing modules for confirmation.
- An e-mail message is dispatched to a predetermined list of recipients, informing them that a fire has been detected on the bridge. The list of recipients can be easily changed to add or delete recipients.

- Each of the three sensing-system modules is given instructions to store buffer data and record live data for a specified amount of time (typically 5 minutes).
- Once the specified buffer period time has elapsed, all three sensing-system modules are instructed to save both the data from the buffer and the live recording to a specified location on the onsite computer.

In the event that the infrared camera detects a person onsite at the bridge during late night or early morning hours, the infrared module sends a person event status report to the main module. At this point, the following events occur:

- The IR module then instructs the camera to record an infrared picture at specified intervals (10 seconds) for a specified period of time (1 min) (Fig. 9). These pictures are then saved to a specified location on the onsite computer.
- An e-mail message is dispatched to a predetermined list of recipients at the BEC, informing the BEC that a person has been detected at the bridge. The list of recipients can easily be changed to add or delete recipients.

In the event that a fault status report is received by the main component, an e-mail is sent to a predetermined list of recipients at the BEC, informing them that a fault has been detected within the flame detector. The flame detector can be powered off and restarted remotely.

Sensing Module Programming

Each of the sensing modules can be configured and adjusted to fine-tune the sensitivity of their respective event triggers and reduce the number of false alarms. The logic used to develop the programming for each sensing module is presented in the following paragraphs.

The flame detector module has a number of configurable parameters that relate to detection of a fire or a fault. A positive voltage reading must be received for the flame detector module to send a fire or fault event report to the main module.

The administrator sets the minimum voltage value before a positive value is recorded for either a fire or fault. A minimum number of consecutive positives—typically five— must occur before the event report is transmitted to the main module.

The flame detector is polled on a set time interval (typically once per second) by the flame detector component to determine the voltage readings.

The interrogator module can be adjusted to control the threshold temperature that must be recorded prior to issuing a fire event report to the main module. A specified number of consecutive readings above the threshold temperature (typically 5) must be recorded before the fire event is reported to the main module.

The interrogator module includes a data buffer that stores the temperatures recorded by each FOS for a preceding set amount of time. It is these data that are saved after a fire occurs. An initial temperature and corresponding center wavelength for each FOS was recorded at the time of installation. The recorded center wavelength values are converted to temperature readings as specified by the manufacturer and compared with the initial values. The rate of temperature change at a particular sensor must exceed a specified threshold value prior to a fire event report being transmitted to the main module.

The final sensing-system component is the IR camera. The first parameter set in the camera component is how often the camera will send data to the camera component. The data that are sent by the camera can be either a picture or temperatures associated with a configurable box or spot in the camera's field of view. The administrator can create up to four boxes and four spots of interest in the camera's field of view.

The temperature data corresponding to these boxes are the average, minimum, and maximum detected, and the data corresponding to the spots are the detected temperature. The camera component has a data buffer that stores the data from the camera for a predetermined set amount of time. In order for the camera component to issue a fire alarm, the maximum temperature in a box of interest must exceed a set threshold temperature, and this condition must be met for a specified number of consecutive readings. The camera component also needs to detect the presence of humans during restricted times. Thus, the restricted times need to be set to prevent false alarms from occurring during regular visiting hours.

If a person is detected during the restricted times or a fire event is issued, the camera component will take pictures at a set time interval for a set time period.

The camera module uses two independent tests to detect a person (note that people can be distinguished from animals through the heat signatures associated with their hands and head) within the field of view:

- **Using a threshold temperature as an indicator.** When the camera module records a maximum temperature reading exceeding the threshold a set number of consecutive times, a person event status report will be sent to the main module.
- **Using a temperature differential as an indicator.** In this test, two rectangular regions are defined within the field of view. A temperature-control box is defined somewhere in the camera's field of view where people are highly unlikely to be present. A temperature test box is defined and contains the regions most likely to contain a human. The surface temperature from the control box or spot is compared with the temperature in the test box. If the difference between the two temperatures exceeds a specified value, a person event report is sent to the main module.

In most cases, the temperature differential method appears to give more reliable results and reduces the potential number of false alarms. Onsite test results will be presented in the System Evaluation section. Examples of events recorded by the onsite monitoring system are in the Appendix.

Remote Monitoring

A local internet service provider (ISP) in Madison County was hired to enable the BEC to remotely monitor the system installed at the bridge. The research team was able to take advantage of the fact that the local businessman who owned the ISP was the same individual who has received a grant to install Web-based security cameras at all covered bridge sites in Madison County. Since the Web-based cameras were connected to the internet via a wireless access point (WAP) near the bridge, it was very convenient to use this same connection for data transmission from the onsite monitoring system. In fact, the utility pole where the infrared camera was to be installed is the same location where the Web-based camera and the WAP were installed. An IP address was assigned by the local ISP to the onsite computer. From the WAP, a signal is transmitted to the nearest wireless transceiver within the local ISP network to the ISP's main computer server, which is connected to the internet through telephone lines.

The monitoring system was designed such that it can be remotely accessed via a desktop connection to connect to the IP address assigned to the onsite computer. This remote access allows the user to see and control the onsite computer from anywhere in the world.

Figure 10. Cabinet installed near utility pole.

Onsite Installation of Components

The selected sensors and computer equipment were installed at the Cedar Bridge site during the summer and fall of 2005. The following briefly summarizes the installation and connection of these components.

Weather-Tight Cabinets

Two weather-tight, steel, locking cabinets, formerly used to house traffic signal equipment, were installed at the bridge site to protect the instrumentation and monitoring equipment. These cabinets, and the equipment they contained, are as follows:

- Located beneath the south approach span of the bridge—containing the FOS interrogator, wireless bridge, and the 5-volt power supply and signal conditioners for the flame detector.
- Located near the base of the utility pole—containing the onsite computer and wireless router (Fig. 10).

The cabinets were supported on small, cast-in-place concrete pads and supplied with electrical power through an agreement between Madison County and the Winterset Municipal Utility. The Madison County engineer supplied equipment and labor to excavate a small trench to permit shared electrical power between the cabinets. Unfortunately, the distance between the cabinets prevented the direct connection of computer equipment and components using the same trench. Both enclosures were equipped with a heater and a fan to help control the temperatures within the cabinets. The two enclosures were purchased for approximately $800 each and were installed in summer 2005.

Flame Detector

The initial location selected for the flame detector was within the roof truss members directly above the south portal of the bridge. This location provided a very uniform monitoring coverage of the south half of the bridge deck, but unfortunately, the view from this point was largely obstructed by framing members of the bridge roof system (Figures 11 and 12).

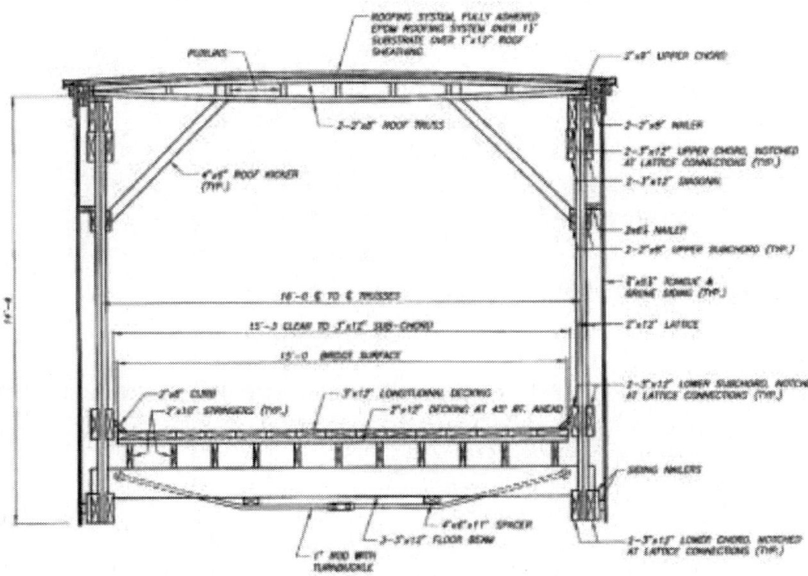

Figure 11. Section view of bridge showing original and revised location of flame detector.

The flame detector was subsequently relocated to a location near the southwest corner of the roof of the covered portion of the bridge (Fig. 13). All the wiring was concealed between the truss framing and outer sheathing of the bridge. The flame detector was covered with a small wooden box to help preserve the aesthetic appearance of the bridge. Note that even in the alternate location, the flame detector view was still slightly compromised by obstructions from roof kicker members, and multiple flame detectors would be required for true 100% coverage (Figure 13).

Figure 12. Original location for flame detector above south portal.

Figure 13. Flame detector installed near southwest portal of bridge.

Fiber Optic Sensors

A total of 12 fiber optic sensors were installed in two series of six and were located along the outer edge of the roadway at the south end of the bridge (Figure 14). The sensors were attached to the back side of the truss framing and hidden from view to the greatest extent possible (Figure 15).

Insulated staples were hand-hammered to attach the connecting patch cords to the truss framing.

Figure 14. View of flame detector slightly obscured by roof kicker members.

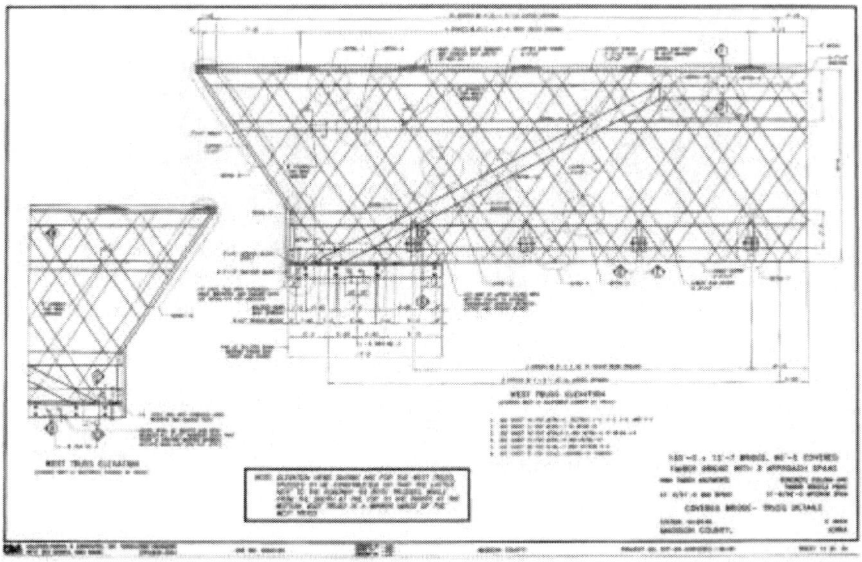

Figure 15. Truss elevation showing fiber optic sensor locations.

The staples were selected to provide a loose fit around the fiber to ensure that the sensors were not rigidly attached to the framing and thus were able to measure thermal changes and not mechanical strains. Fiber optic patch cords were run from the two closest FOS along the underside of the bridge, into the enclosure and connected to the interrogator. Although precautions were taken to conceal the wires during installation, one series of six sensors (along the east truss) were disabled by an unknown person during the early months of 2006 (Figures 16 and 17). The glass fiber within each patch cord is protected by a thin layer of Kevlar within the outer jacket, so it takes considerable effort to cut one of these fibers, which makes it unlikely that this damage was caused by any accidental event. Loss of the fiber was reported as an "event."

Figure 16. Installation of fiber optic sensors behind truss members.

Figure 17. Fiber optic sensor series damaged by possible vandalism.

Infrared Camera

The camera enclosure was installed approximately 25 ft above the ground on the previously mentioned utility pole (Figure 18). The infrared camera was aligned to view the entire bridge site including the approach roadway on both ends of the bridge. Unfortunately from this vantage point, a single mature tree created a significant viewing obstruction during the growing season. A request was made of the county parks department to selectively prune some of the branches.

Numerous pieces of hardware were used in the installation of the system on the bridge. These pieces of hardware were used for connections, camouflage, and protection of wiring. The total cost for the miscellaneous pieces of hardware was approximately $2,000.

Figure 18. Infrared camera (bottom camera) and web-based video camera (top camera) attached to utility pole.

SYSTEM EVALUATION

Each of the sensor components was evaluated individually for effectiveness and reliability in the laboratory located on the ISU campus as well as following installation in the field. The results of these evaluations are presented here.

The recognized industry standard used for fire detection equipment uses a gasoline fire covering an area of 1 ft^2 (Factory Mutual 1994). A gasoline fire was considered too volatile for use on an historic covered bridge; therefore, denatured ethyl alcohol was used for both the laboratory and field evaluations in place of gasoline. A galvanized metal pan measuring 12 in. diameter (0.79 ft^2) was used to contain the alcohol during each test.

Laboratory Evaluation

The individual sensor components were tested for sensitivity and reliability prior to their assembly into a complete system. The tests performed for this current study used three sources of heat and flame including

- Butane lighter (tested at 3-in. increments from sensor)
- Disposable propane torch (tested at 6-in. increments from sensor)
- Alcohol pan fire as described above (tested at 1-ft increments from sensor)

In each test, the heat source was positioned near the sensor and data collected until a steady-state was reached. The heat source was then moved to the next incremental distance from the sensor and the process repeated. The goal with this testing was to determine the distance and associated rate of detection at which a fire could be detected by each sensing system.

Fiber Optic Sensors

A series of tests was performed to determine the response of the FOS to the heat generated by each source. The sensor was positioned at a range of distances up to 6 ft from the heat sources. The results from these tests are shown in Figures 19 and 20.

From these test results, it is evident that the fiber optic sensor was not able to record any significant temperature changes from the pan fire beyond approximately 2 ft.

However, since the installation location on the actual bridge will be near the deck level and hidden behind the truss, a fire very likely could be started in this area (and within a few feet of a sensor).

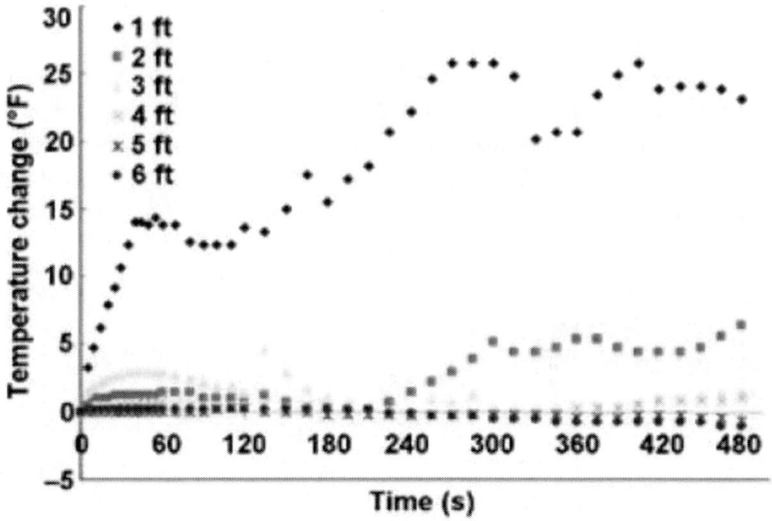

Figure 19. Fiber optic sensor sensitivity to pan fire.

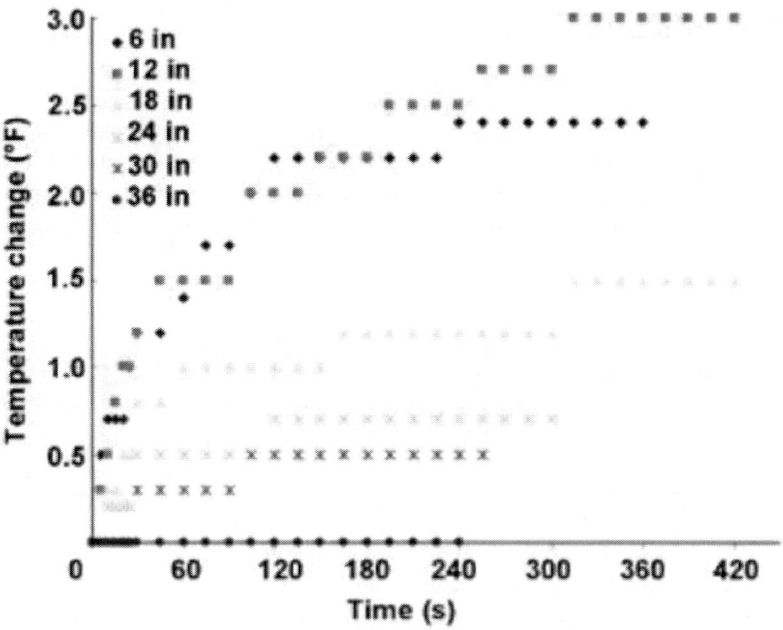

Figure 20. Fiber optic sensor sensitivity to propane torch.

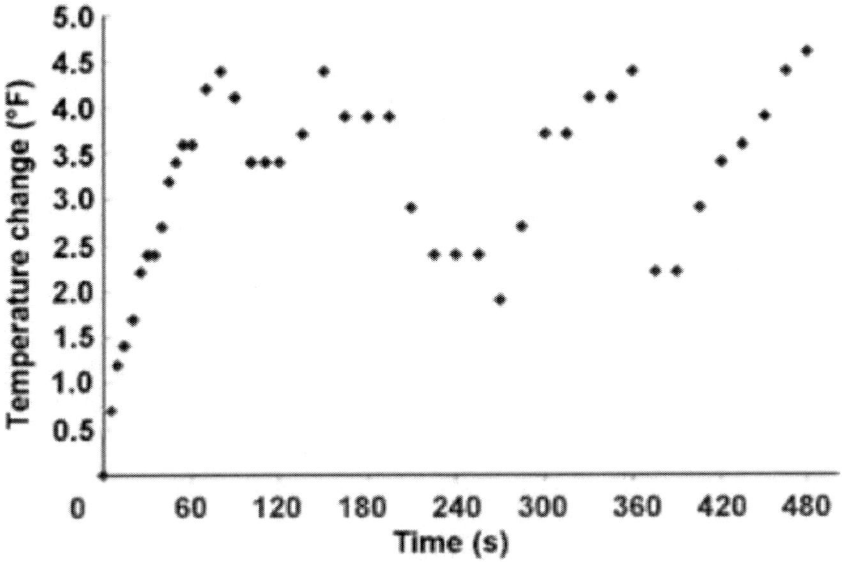

Figure 21. Fiber optic sensor sensitivity to pan fire with wood shielding.

In the case of the propane torch, the maximum temperature change recorded was approximately 3 degrees. We thought that perhaps 5 degrees is the minimum threshold of localized temperature change that could be reliably used to trigger a fire alarm, and thus no further testing was performed using smaller heat sources, for instance, the butane lighter.

To test the ability of a FOS to detect temperature changes when separated from the heat source by a layer of wood, a similar test was performed using a nominal 1- in.-thick layer of cedar wood as a screen. For this test, the fire was located 1 ft from the sensor. The results from this test are shown in Figure 21. The maximum recorded temperature change was approximately 4.5 degrees, indicating that a FOS would not be able to reliably detect a fire on the opposite side of a timber member or exterior sheathing.

Infrared Camera

Two series of laboratory tests were performed using the infrared camera using different fire sources and at varying distances from the sensor. Independent measurement of fire temperature was made using a handheld infrared thermometer. In each test, five infrared images of the flame were captured at 15–20 second intervals. The following heat sources were used to evaluate the infrared camera sensing system:

- A butane lighter located at 1-ft increments up to 20 ft from the camera
- Pan fire located at 10-ft increments up to 250 ft from the camera

Subsequent analysis of the infrared images from both the lighter tests and the pan fire tests indicate that the IR camera is able to detect temperatures very close to the actual fire temperatures almost immediately. A sample of the IR camera images recorded is presented in Figures 22 and 23.

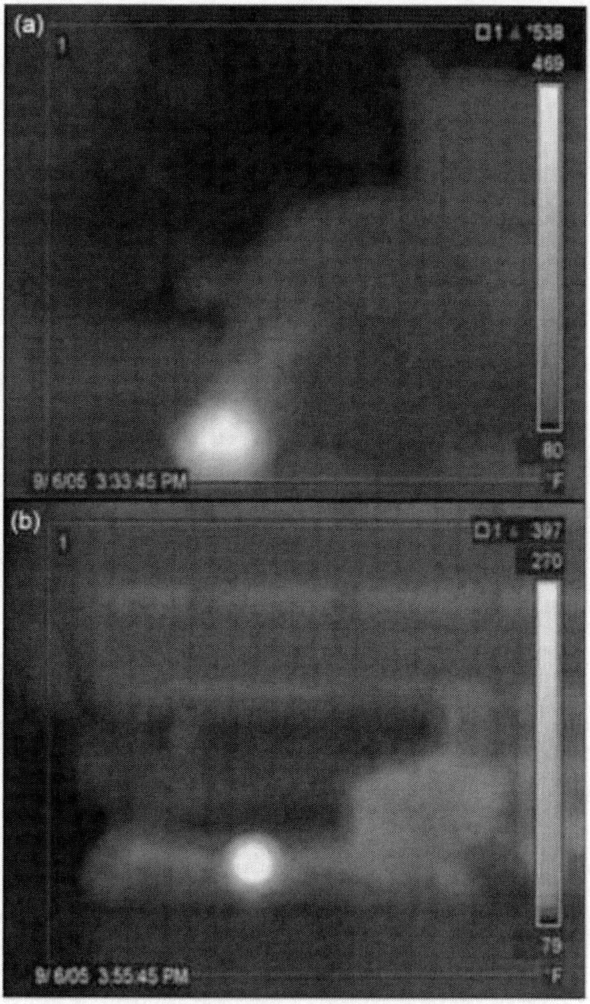

Figure 22. Infrared photo—pan fire with a) fire located 30 ft from camera and b) fire located 100 ft from camera.

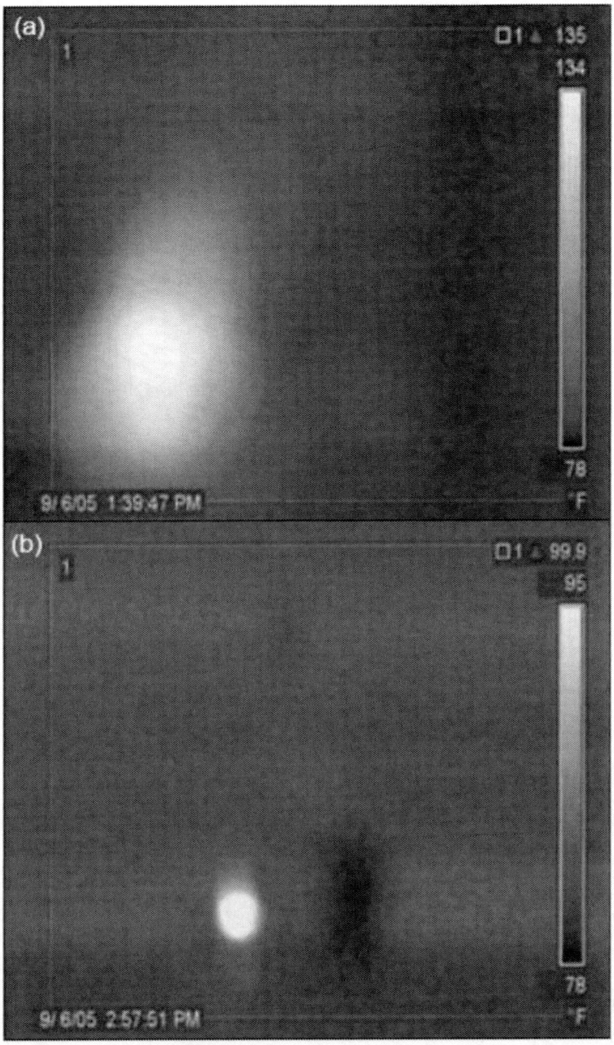

Figure 23. Infrared photo—butane lighter with a) fire located 2 ft from camera and b) fire located 20 ft from camera.

Flame Detector

A series of laboratory tests was performed to determine how quickly the flame detector could reliably detect a fire of various sizes at a range of distances. An initial series of tests were performed using the 12-in. pan fire, located at a distances of 10–80 ft from the source. The laboratory setup for this test is shown in Figure 24.

Figure 24. Laboratory setup for flame detector tests.

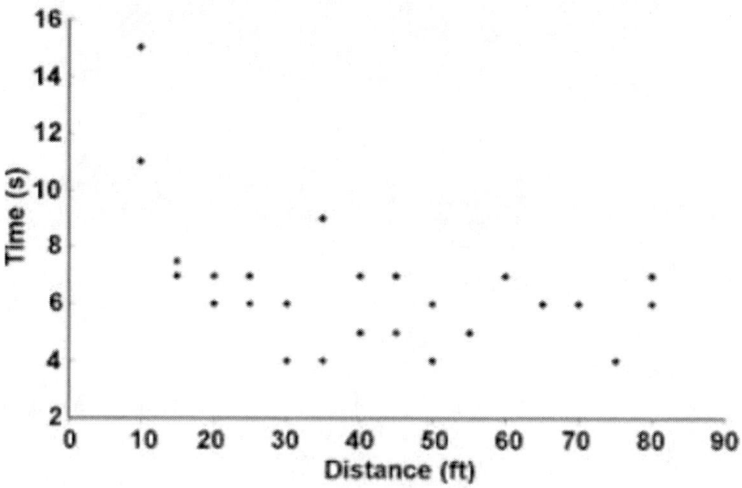

Figure 25. Flame detector sensitivity to pan fire.

The manufacturer claims that the sensor can reliably detect a 1 ft^2 pan fire at a distance of 15 ft in 5 seconds. Note that although the test results indicate detection occurs at approximately 7 seconds, the pan fire used for this test covers approximately 22% less area than the reference fire used by the manufacturer. The results of these tests are presented in Figure 25. An additional series of tests was performed using a butane lighter as the flame source. In this test, the flame detector was located at distances from 3–24 in. from the source. Beyond this distance, the sensor could no longer detect the flame within 5 min. The results of these tests are shown in Figure 26.

Onsite Evaluation

A series of onsite tests was performed to evaluate the individual component's effectiveness in detecting fire events. In each of these fire tests, a 12-in. diameter metal pan was used to contain the fire and a small four-wheeled cart was used to allow the fire to be easily moved on the bridge deck without the need to extinguish the fire prior to each test. All fire tests were performed with the knowledge and consent of the local fire and law enforcement authorities. In addition to extreme caution used by the research team when using flammable materials near the bridge, a fully-charged fire extinguisher was available at all times during these tests. In addition to fire testing, the infrared camera was tested for its ability to detect a person moving about at the bridge site during selected hours of the day.

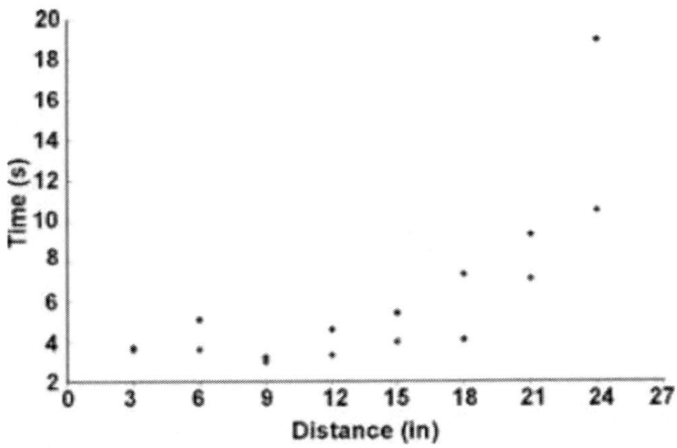

Figure 26. Flame detector sensitivity to butane lighter.

Fiber Optic Sensors

The FOS were evaluated for their ability to detect a standard pan fire during both daylight and darkness. The pan fire was located as near to the bridge sheathing as deemed safe without risking the possibility of unintentional damage to the bridge. The results of these tests indicate that a pan fire could be detected within 5–8 seconds when the flame source was located within a few feet of a sensor.

Figure 27. Pan fire test on south bridge approach span.

Infrared Camera

The infrared camera was subjected to fire tests during both daytime and nighttime hours. These tests were performed with the standard pan fire positioned at several locations on the south approach span of the bridge (Fig. 27). In all cases, the fire was detected within less than 10 seconds in all tests and in less than 5 seconds in over 75% of the tests.

In addition to the fire tests, a series of intruder tests was performed that consisted of a person moving about on the bridge site during the hours of darkness. A series of photos (Fig. 28) presents the infrared images captured during both fire and intruder tests. The image of a person is clearly visible in a), which indicates that the IR camera is fully capable of this type of detection.

Flame Detector

A series of pan fire tests were performed to determine the area of the bridge which could be reliably monitored by the flame detector. A rectangular grid, measuring 4 ft by 4 ft, was used to locate the pan fire prior to each test. These tests were performed during both daylight and dark hours.

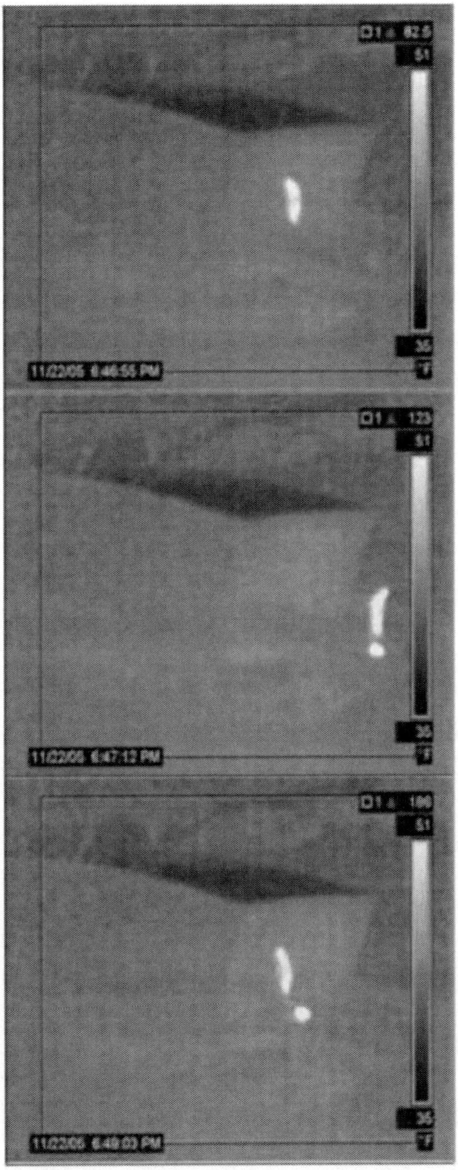

Figure 28. Infrared photos—onsite system evaluation of a) person only, b) person and initial fire, c) person and fire after 2 min.

In fact, the detection of a fire was not limited by the distance from the flame source to the sensor. Instead, the primary impediment to reliable detection using the flame detector was line-of-sight interference from bridge roof framing members. In all cases where interference was not present, the flame detector was able to detect a fire within 5–7 seconds.

Conclusion and Recommendations

The instrumentation and monitoring of the Madison County Cedar Bridge presented in this report has provided an excellent first step in the development of security and fire protection monitoring of important historic resources.

A total of three sensing systems were installed and tested both in the laboratory as well as at the actual bridge site. These systems included the following:

- Infrared camera mounted on utility pole to monitor the bridge and approach roadways
- Industrial flame detector mounted near the bridge portal to monitor the bridge interior
- Fiber optic sensors mounted near the timber bridge deck to monitor ambient temperature changes

The three sensor systems were connected via a wired and wireless network to a computer running custom-designed software which collects data from each of the sensors, determines whether a fire or intruder event has taken place, and notifies interested parties. The three systems were evaluated on the basis of cost and effectiveness. Based on the ease of installation, reliability, and relatively low cost of the industrial flame detector used on this project, these units appear to offer the greatest potential benefit to bridge owners at detecting fires in similar installations. The primary impediment to the reliable use of similar flame detector units is line-of-sight interference from bridge framing members. Future installations should carefully consider the placement of these sensors or provide a number of sensors with differing viewpoints to eliminate these interferences. The infrared camera proved to perform well at detecting the presence of persons at the bridge site. This sensing system, when properly placed, can provide an early warning to authorities of persons at a location where they should not be. The research work performed during this project generated a considerable amount of media attention in the local, regional, and national press. This increased visibility may in some small way have improved the public recognition of the critical need to protect vital historical resources. Future developments that could be made to improve security monitoring of covered bridges:

- Application of solar power systems which can provide sufficient power for monitoring without overly affecting the aesthetic appearance of historic bridge sites
- Development and installation of a wide area monitoring network, which would permit an owner-agency to continuously monitor all of the covered bridges under their jurisdiction.

REFERENCES

Cho, A. 2005. Cameras and sensors will patrol New York rail, bridges. *Engineering News-Record.* September 5; 16.

Crossett, J.; Rhodes, B. 2005. Protecting America's roads, bridges, and tunnels: the role of state DOTs in homeland security. Report RP–PAR–1. Washington, DC: American Association of State Highway and Transportation Officials.

Factory Mutual. 1994. Flame radiation detectors for automatic fire alarm signaling. FM 3260. Norwood, MA: Factory Mutual.

Roberts, J.E.; Kulicki, J.M.; Beranek, D.A.; Englot, J.M.; Fisher, J.W.; Hungerbeeler, H.; Isenberg, J.; Seible, F.; Stinson, K.; Tang, M.C.; Witt, K. 2003. Recommendations for bridge and tunnel security. Washington, DC: American Association of State Highway and Transportation Officials, Transportation Security Task Force.

Stevens, M.A. 2005. Security essentials. *Buildings*. 99(8): 42–46.

Williamson, E.B.; Winget, D.G. 2005. Risk management and design of critical bridges for terrorist attacks. *Journal of Bridge Engineering.* 10(1): 96–106.

Wolff, T.O. 2003. Proposed TMC response protocol to suspicious activity at highway bridges. Advanced surface transportation systems class. College Station, TX: Texas A&M University: 56–75.

APPENDIX A—EVENTS RECORDED BY ONSITE MONITORING SYSTEM

Nine events were recorded by the system in place at Cedar Bridge (Table A-1, Figures A-1–A9).

Table A–1. Summary of recorded events

Event number	Date	Time	Type	True/ False
1	01/29/2006	2:29:46 a.m.	Intruder	True
2	03/03/2006	10:07:33 p.m.	Intruder	True
3	03/07/2006	10:40:13 p.m.	Intruder	True
4	03/11/2006	9:04:21 p.m.	Intruder	True
5	04/19/2006	2:57:57 p.m.	Fire	False
6	04/27/2006	9:10:14 a.m.	Fire	False
7	05/04/2006	10:27:45 a.m.	Fire	False
8	05/24/2006	7:41:38 a.m.	Fire	False
9	06/16/2006	4:10:32 a.m.	Intruder	False

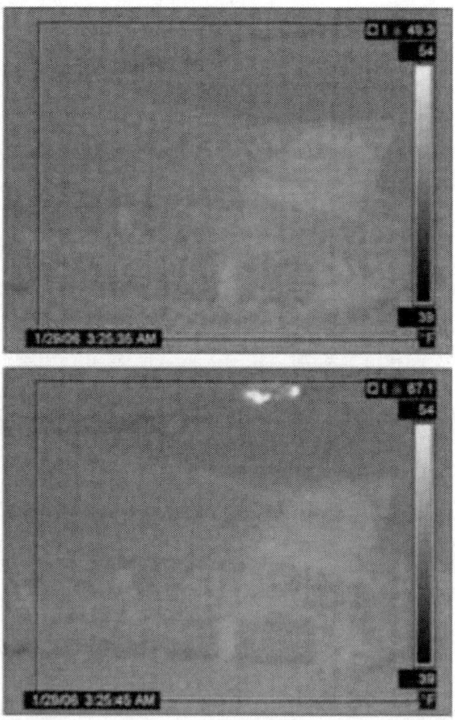

Figure A–1. (Continued).

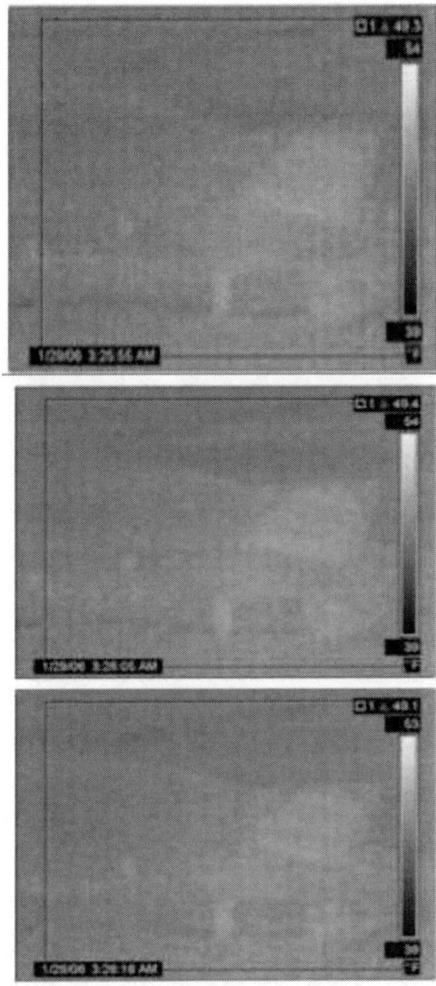

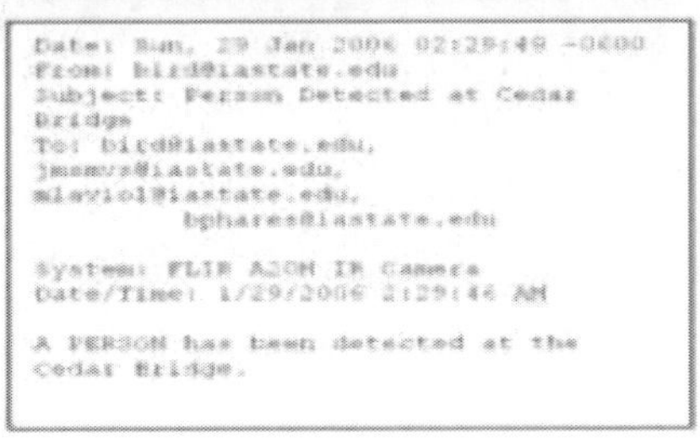

Date: Sun, 29 Jan 2006 02:29:49 -0600
From: bird@iastate.edu
Subject: Person Detected at Cedar Bridge
To: bird@iastate.edu,
jmsmvs@iastate.edu,
mlaviol@iastate.edu,
bphares@iastate.edu

System: FLIR A20M IR Camera
Date/Time: 1/29/2006 2:29:46 AM

A PERSON has been detected at the Cedar Bridge.

Figure A–1. Event 1—Person detected at Cedar Bridge (1/29/06, 2:29 a.m.) a. Image 1, b. Image 2, c. Image 3, d. Image 4, e. Image 5, f. E-mail notification.

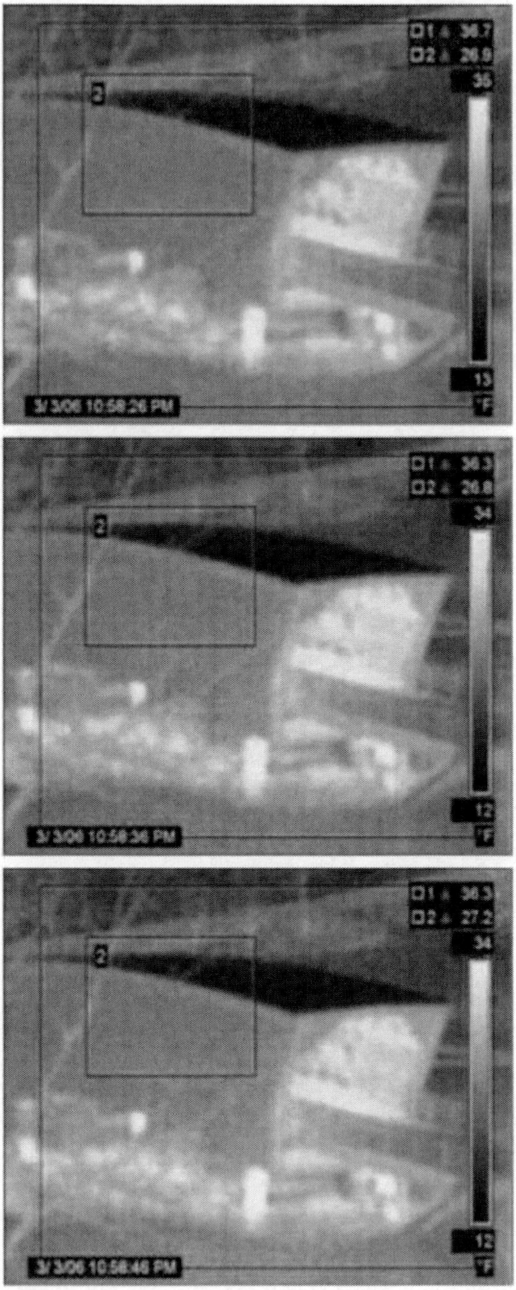

Figure A–2. (Continued).

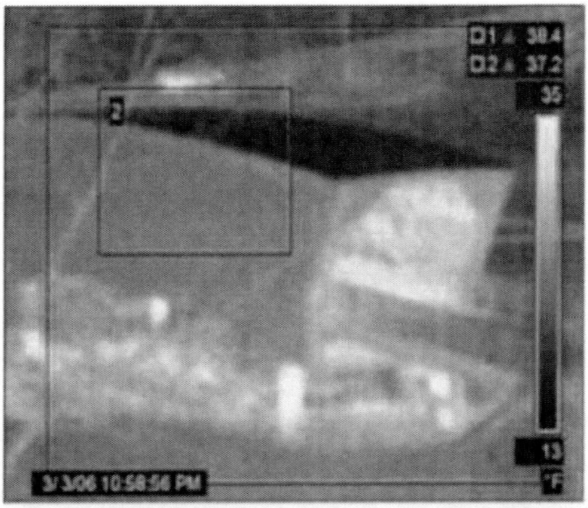

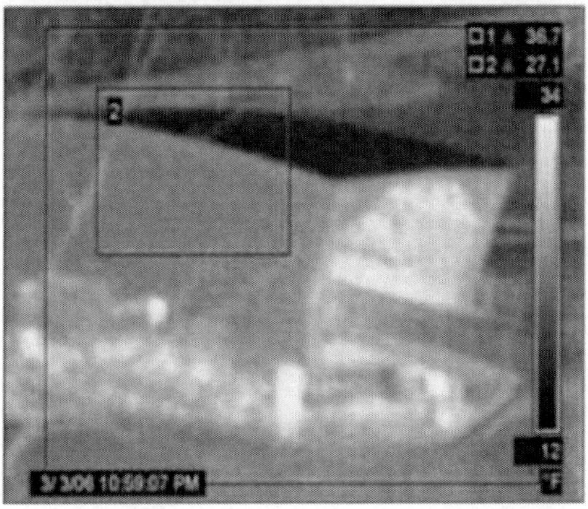

Date: Fri, 3 Mar 2006 22:07:45 -0600
From: bird@iastate.edu
Subject: Person Detected at Cedar Bridge
To: bird@iastate.edu,
jmsmvs@iastate.edu,
mlaviol@iastate.edu,
bphares@iastate.edu

System: FLIR A20M IR Camera
Date/Time: 3/3/2006 10:07:33 PM

A PERSON has been detected at the Cedar Bridge.

Figure A–2. Event 2—Person detected at Cedar Bridge (3/3/06, 10:07 p.m.) a. Image 1, b. Image 2, c. Image 3, d. Image 4, e. Image 5, f. E-mail notification.

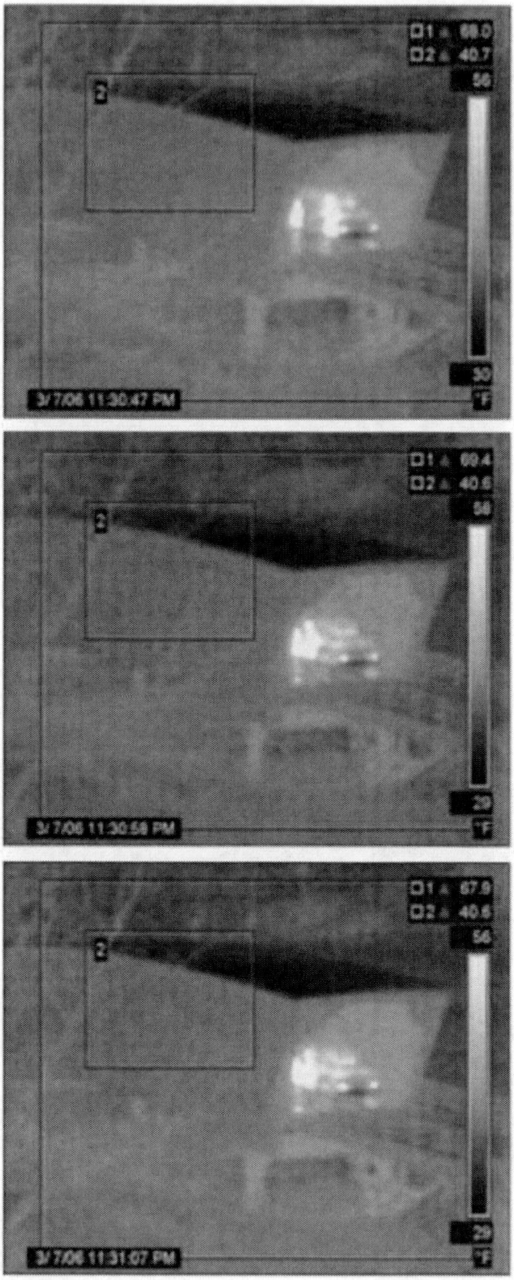

Figure A–3. (Continued).

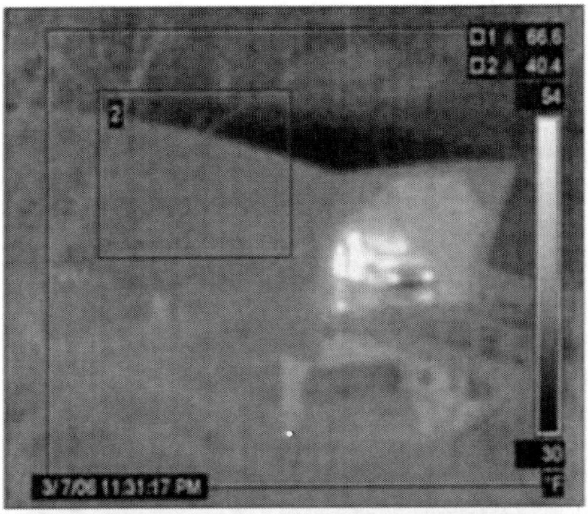

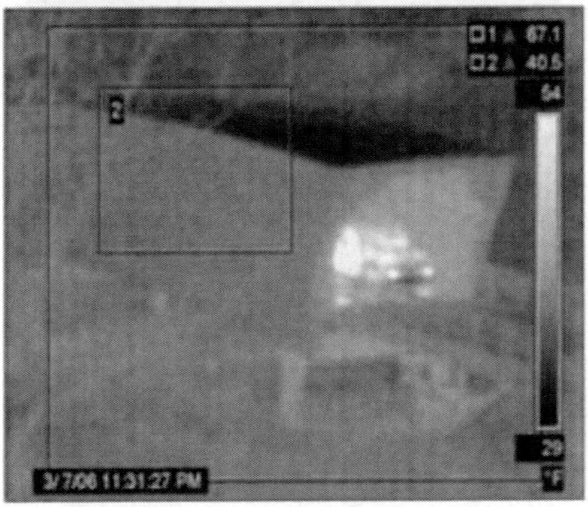


```
Date: Tue, 7 Mar 2006 22:40:17 -0600
From: bird@iastate.edu
Subject: Person Detected at Cedar
Bridge
To: bird@iastate.edu,
jmzmva@iastate.edu,
mlaviol@iastate.edu,
          bphares@iastate.edu

System: FLIR A20M IR Camera
Date/Time: 3/7/2006 10:40:13 PM

A PERSON has been detected at the
Cedar Bridge.
```

Figure A–3. Event 3—Person detected at Cedar Bridge (3/7/06, 10:40 p.m.) a. Image 1, b. Image 2, c. Image 3, d. Image 4, e. Image 5, f. E-mail notification.

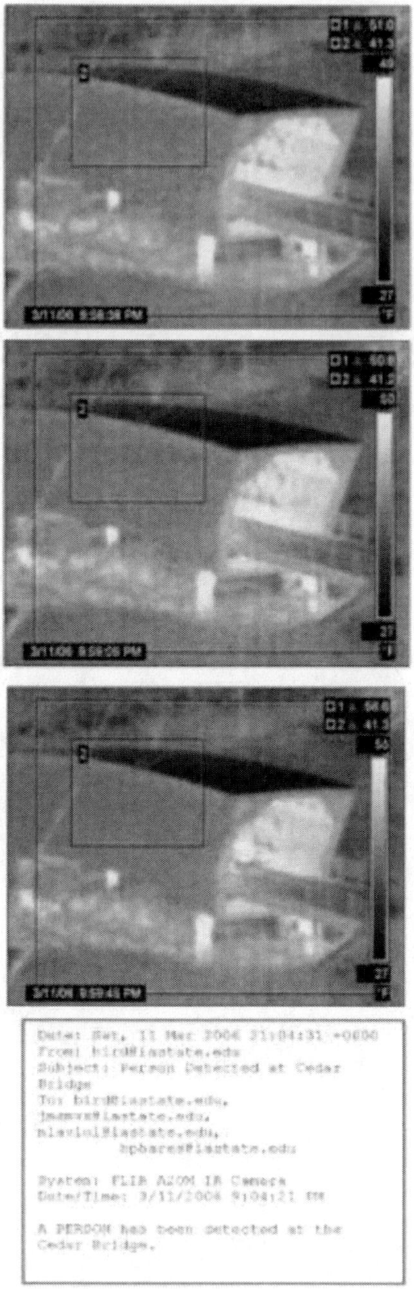

Figure A–4. Event 4 – Person detected at Cedar Bridge (3/11/06, 9:04 p.m.) a. Image 1, b. Image 2, c. Image 3, d. E-mail notification.

Figure A–5. (Continued).

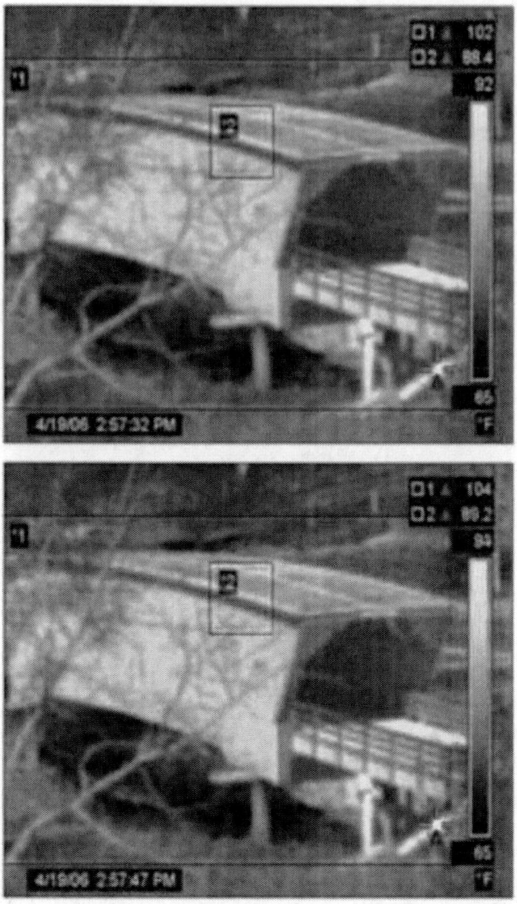

Date: Wed, 19 Apr 2006 15:06:56 -0500
From: bird@iastate.edu
Subject: Fire Detected at Cedar Bridge
To: bird@iastate.edu,
jmsmvs@iastate.edu,
mlaviol@iastate.edu,
bphares@iastate.edu

System: FLIR A20M IR Camera
Date/Time: 4/19/2006 3:06:47 PM

A FIRE has been detected at the Cedar Bridge.

Figure A–5. Event 5—Fire detected at Cedar Bridge (4/19/06, 2:57 p.m.) a. Image 1, b. Image 2, c. Image 3, d. Image 4, e. Image 5, f. E-mail notification.

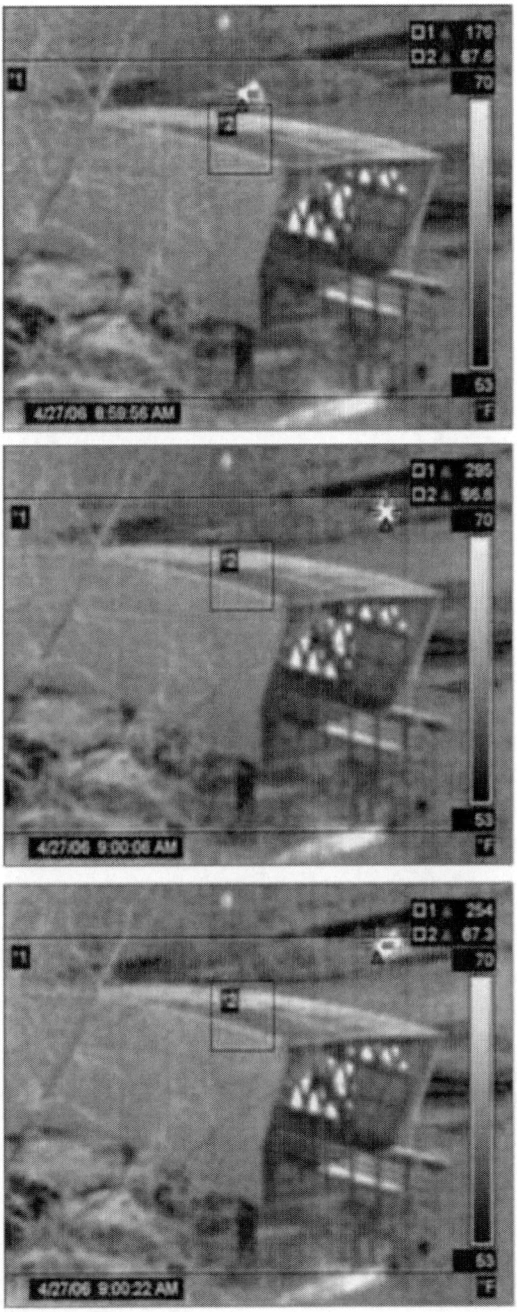

Figure A–6. (Continued).

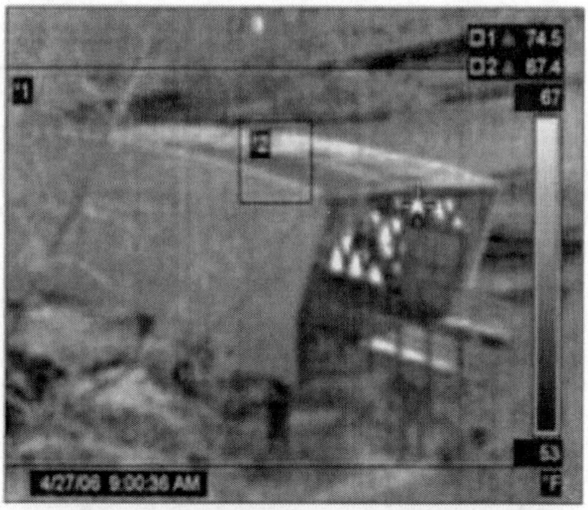

```
Date: Thu, 27 Apr 2006 09:10:33 -0500
From: bird@iastate.edu
Subject: Fire Detected at Cedar Bridge
To: bird@iastate.edu,
jmsmvs@iastate.edu,
mlaviol@iastate.edu,
          bphares@iastate.edu

System: FLIR A20M IR Camera
Date/Time: 4/27/2006 9:10:14 AM

A FIRE has been detected at the Cedar
Bridge.
```

Figure A–6. Event 6—Fire detected at Cedar Bridge (4/27/06, 9:10 a.m.) a. Image 1, b. Image 2, c. Image 3, d. Image 4, e. Image 5, f. E-mail notification.

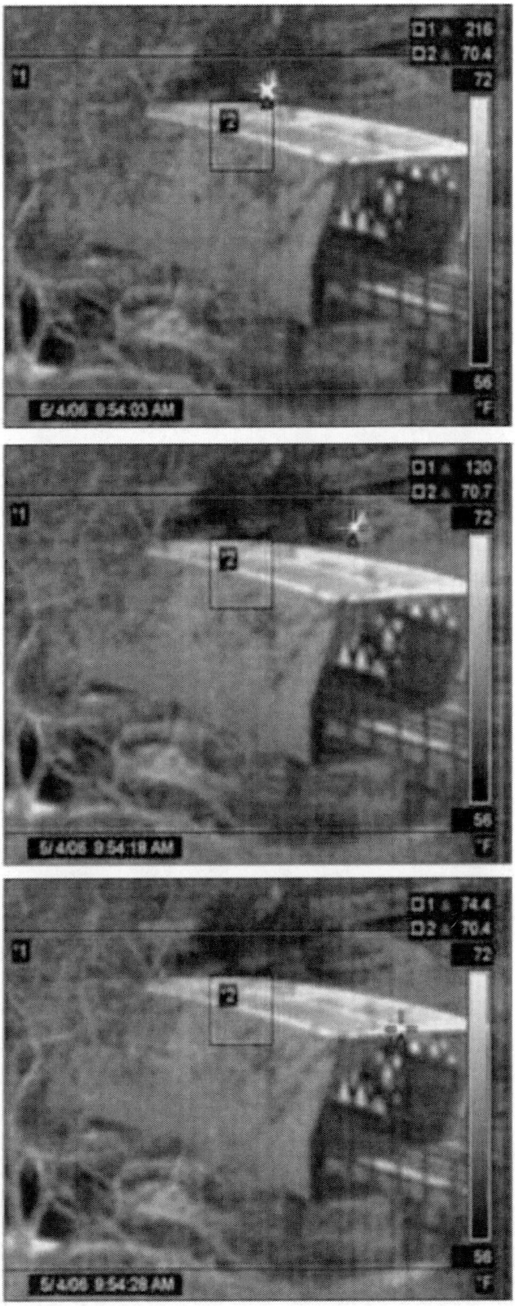

Figure A–7. (Continued).

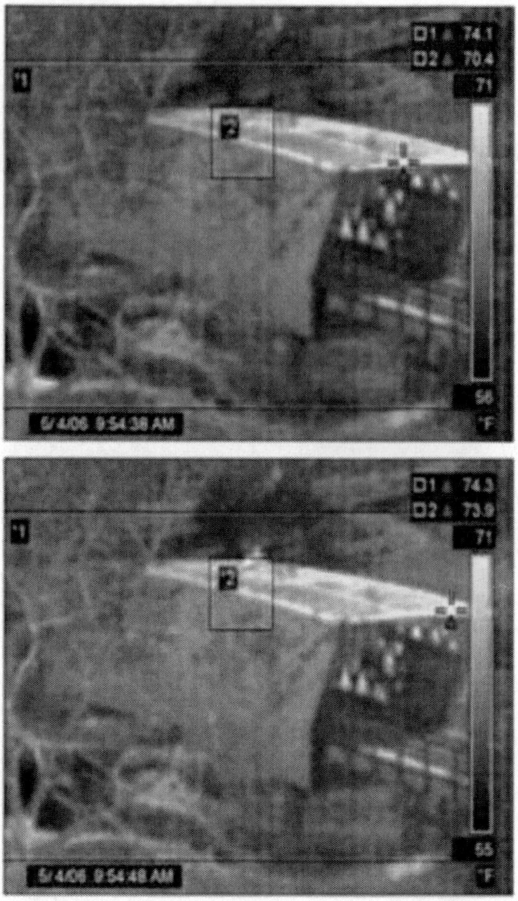

Date: Thu, 4 May 2006 10:27:53 -0500
From: bird@iastate.edu
Subject: Fire Detected at Cedar Bridge
To: bird@iastate.edu,
jmsmvs@iastate.edu,
mlaviol@iastate.edu,
bphares@iastate.edu

System: FLIR A20M IR Camera
Date/Time: 5/4/2006 10:27:45 AM

A FIRE has been detected at the Cedar Bridge.

Figure A–7. Event 7—Fire detected at Cedar Bridge (5/4/06, 10:27 a.m.) a. Image 1, b. Image 2, c. Image 3, d. Image 4, e. Image 5, f. E-mail notification.

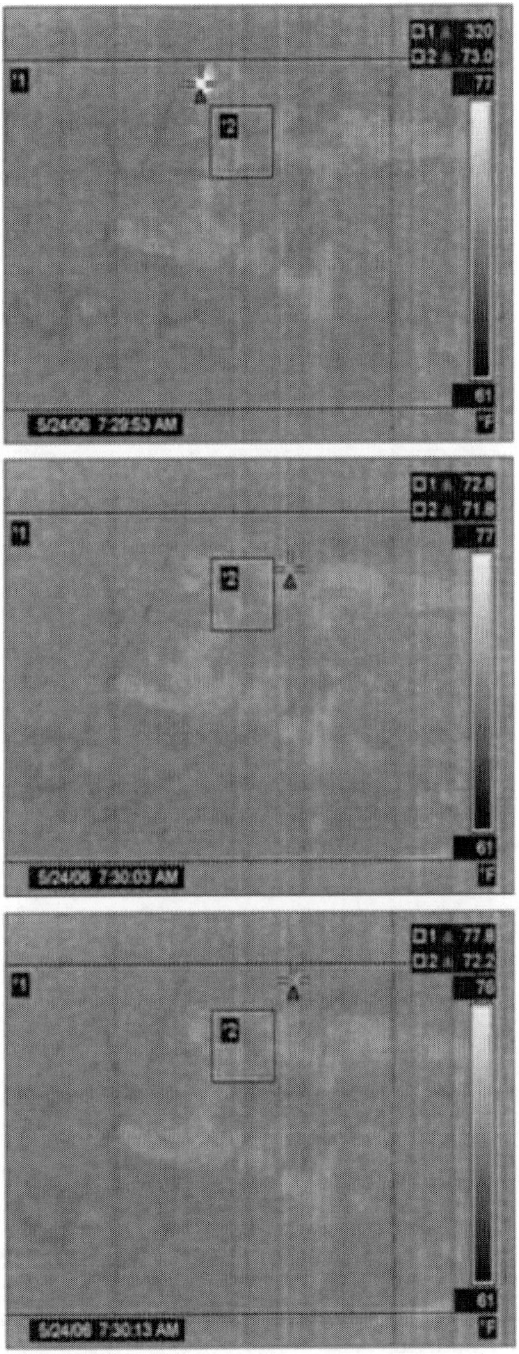

Figure A-8. (Continued).

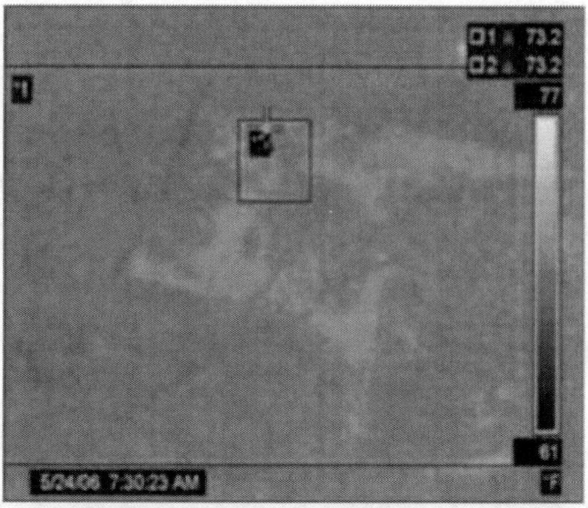

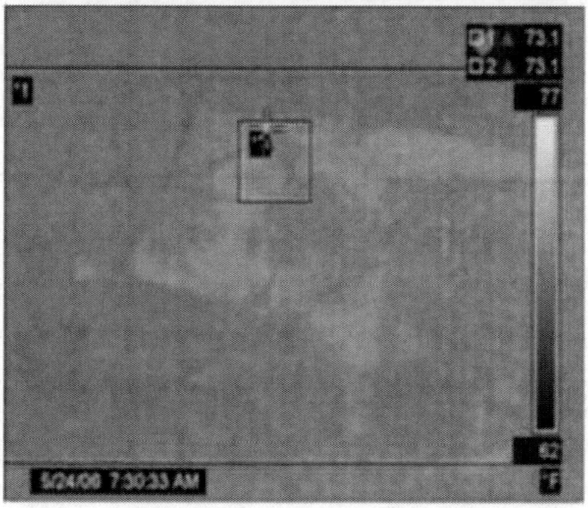


```
Date: Wed, 24 May 2006 07:41:40 -0500
From: bird@iastate.edu
Subject: Fire Detected at Cedar Bridge
To: bird@iastate.edu,
jmamva@iastate.edu,
mlaviol@iastate.edu,
          bphares@iastate.edu

System: FLIR A20M IR Camera
Date/Time: 5/24/2006 7:41:38 AM

A FIRE has been detected at the Cedar
Bridge.
```

Figure A-8. Event 8—Fire detected at Cedar Bridge (5/24/06, 7:41 a.m.) a. Image 1, b. Image 2, c. Image 3, d. Image 4, e. Image 5, f. E-mail notification.

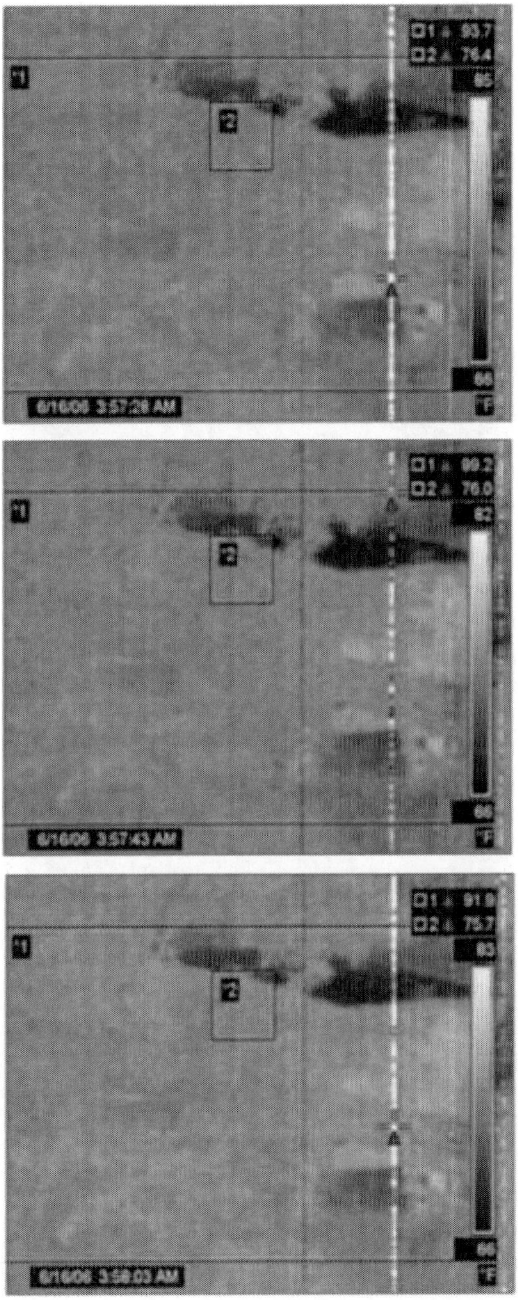

Figure A–9. (Continued).

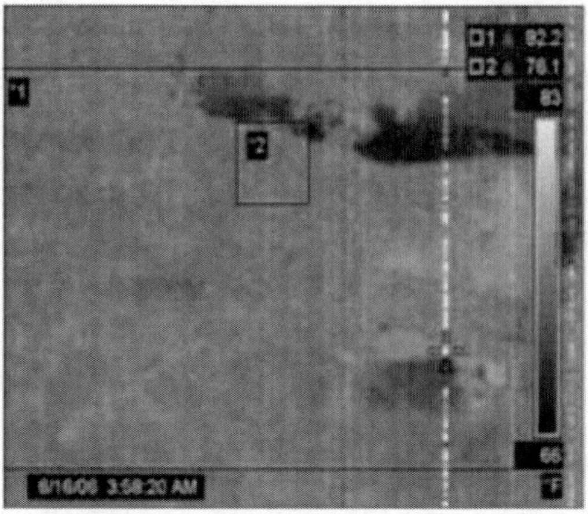

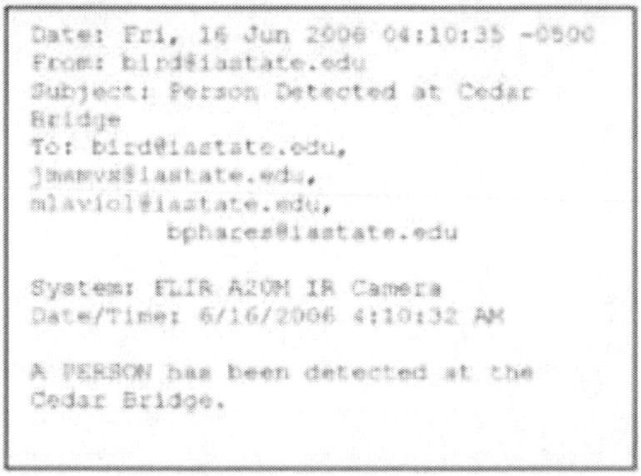
Date: Fri, 16 Jun 2006 04:10:35 -0500
From: bird@iastate.edu
Subject: Person Detected at Cedar Bridge
To: bird@iastate.edu,
jmamvs@iastate.edu,
mlaviol@iastate.edu,
bphares@iastate.edu

System: FLIR A20M IR Camera
Date/Time: 6/16/2006 4:10:32 AM

A PERSON has been detected at the Cedar Bridge.

Figure A–9. Event 9—Person detected at Cedar Bridge (6/16/06, 4:10 a.m.) a. Image 1, b. Image 2, c. Image 3, d. Image 4, e. E-mail notification.

INDEX

D

E

F

L

M

N

O

P

R

S

T

U

V

W